THE REMAKING OF TELEVISION NEW ZEALAND

THE REMAKING OF TELEVISION NEW ZEALAND

1984–1992

Barry Spicer, Michael Powell & David Emanuel

AUCKLAND UNIVERSITY PRESS
in association with the
BROADCASTING HISTORY TRUST

To our families with appreciation for their support.

First published 1996
AUCKLAND UNIVERSITY PRESS
University of Auckland
Private Bag 92019
Auckland

ISBN 1 86940 151 4

Typeset by Auckland University Press
Printed by GP Print, Wellington

CONTENTS

LIST OF FIGURES AND TABLES

LIST OF ABBREVIATIONS

ABC	Australian Broadcasting Commission
BBC	British Broadcasting Commission
BCL	Broadcast Communications Limited
BCNZ	Broadcasting Corporation of New Zealand
CLEAR	Clear Communications Limited
DCF	Discounted cash flow
GDP	Gross domestic product
HDTV	High-definition television
MAC	Ministerial Advisory Committee
PAYE	Pay as you earn
PBF	Public broadcasting fee
PSA	Public Service Association
NZBC	New Zealand Broadcasting Commission
RoA	Return on Assets
RoE	Return on Equity (shareholders' funds)
RNZ	Radio New Zealand
SBU	Strategic Business Unit
SOE	State-owned enterprise
SCI	Statement of corporate intent
SPP	South Pacific Pictures
TCI	TeleCommunications Inc.
TVNZ	Television New Zealand
TV3	TV3 Networks

Preface

This book is about the remaking of Television New Zealand into a commercially successful state-owned enterprise and the management and organisational upheaval experienced by the organisation as a result. The study is rich in detail and reveals how TVNZ Ltd made significant changes to its strategies, structures and management processes as it confronted the threat for the first time of a competing broadcaster entering the industry in which it had for so long held a state-sanctioned monopoly. How TVNZ changed internally to counter this threat and to transform itself into a successful business is the focus of this book. We acknowledge that there are also important cultural and political dimensions to the changes that took place in New Zealand broadcasting and within TVNZ. Although we sketch out some of the main issues involved, we have not attempted to tell the story in cultural or political terms. We leave this for others more skilled in these forms of analysis. Rather we concentrate on telling the story in organisational and management terms from the vantage point of government officials, board members, and managers charged with transforming the state-owned monopoly.

Extensive and detailed studies of institutional and organisational change cannot take place without the support and co-operation of a large number of people. There have been so many individuals who assisted us and gave freely of their time that it is impossible to name them all. However, we would particularly like to acknowledge the support and assistance of Irene Taylor of the New Zealand Treasury who first suggested we include TVNZ in a wider study of the transformation of state-owned enterprises (SOEs) under the State-Owned Enterprises Act 1986. We also thank all the other Treasury officials who supported us in the task as well as the board and management of TVNZ for their co-operation with this study over an extended period of time. Moir Spicer and Leone Hill provided invaluable assistance with transcriptions of interviews and helping us to prepare numerous drafts of this study. We also gratefully acknowledge the support of the Broadcasting History Trust for their financial assistance in publishing our manuscript.

Barry Spicer
Michael Powell
David Emanuel

CHAPTER ONE

Introduction

In a little over a decade, New Zealand has undertaken major reforms of its economy and the operation of its public sector. As all New Zealanders are aware and most of the rest of the world as well,[1] these reforms have involved very significant restructuring of the economy, the deregulation of markets and industries, and the redesign of the governance structures of a host of government enterprises which dominated a number of domestic industries.

One industry to undergo significant structural change under the New Zealand reform programme was television and radio broadcasting. Since the introduction of television in New Zealand in 1960, television broadcasting has been subject to almost continuous public and political pressure. This is reflected in the fact that almost every election in New Zealand since the introduction of television has brought one change or another to the structure of broadcasting. However, decisions made by the Labour government in the late 1980s broke new ground. These included decisions to deregulate broadcasting and the related telecommunications industry, to establish Television New Zealand as a limited liability company under the State-Owned Enterprises Act 1986, and to set up a Broadcasting Commission which would be funded by a broadcasting licence fee and would promote the government's social objectives in broadcasting. Government enterprises engaged in television broadcasting and related telecommunications industries had their legislative protections removed and their markets opened to competition.

Television broadcasting in New Zealand is important for a number of reasons. First, the structure of the television broadcasting industry is significant for economic reasons. As a major medium for ad-

1

vertising and marketing products and services, television broadcasting plays an important role in New Zealand's market economy.

Second, television influences the social and cultural fabric of the nation. There is a strongly held belief in segments of New Zealand society that television broadcasters have special responsibilities not only to inform, educate, and entertain, but also to promote national identity and culture and to meet the needs of minorities. This raises an important question about the extent to which state-owned, as well as privately owned, broadcasters should have special obligations placed on them to 'to serve the public interest' as they carry out their commercial activities.

Third, the international broadcasting environment has gone through a number of profound changes in recent years which have affected the nature, shape, and economics of the industry. Change was inevitable with the introduction of new and affordable delivery systems and television services and with the convergence of transmission technologies bringing new forms of competition for both broadcasters and traditional telecommunications companies around the world. As technological developments rapidly eroded the New Zealand government's ability to protect its state-owned broadcaster from competition from multi-national media and telecommunications companies, major changes to the structure of New Zealand television broadcasting were not long in coming.

Lastly, the structure of the industry and the organisation of broadcasters in New Zealand is important because New Zealand is a small country with a small, scattered population and with a limited capital and income base to support television broadcasting and programme production. It is also a country with challenging geographical features, which make the transmission of television signals to many areas difficult and costly. Working within these limitations has always been a fact of life in New Zealand television.

The confluence of these technological, economic, geographical, political, and social factors added uncertainty and complexity to the problem of how best to structure the broadcasting industry in New Zealand. It makes the restructuring that did take place both interesting and unique. These factors combined to place considerable pressure on the new board and management of the new state-owned enterprise, Television New Zealand Ltd, as they implemented the change that had been mandated by the New Zealand government.

Objectives

This study is set in the context of the radical changes made to the structure of television broadcasting in New Zealand over the period 1984–1992. Its purpose is to examine the organisational and management process of remaking Television New Zealand, a division of the Broadcasting Corporation of New Zealand (BCNZ), into a commercial state-owned enterprise (SOE) under the State-Owned Enterprises Act 1986. We look at these changes from the perspective of TVNZ's managers and those government officials, ministers, and committees involved in making and implementing new broadcasting policies. We focus primarily on organisational and management change rather than on the cultural or political aspects of these changes. However, in the latter case we do attempt to outline some of the main issues involved.

We first set out to understand and evaluate the corporatisation process as it was applied to the television division of the BCNZ. We then analyse the significant organisational and management changes which resulted as the new corporatised entity, TVNZ Ltd, was forced to compete for programmes, audiences, and advertising dollars in a deregulated market environment. The combined interventions of the New Zealand government in deregulating television broadcasting and related telecommunications industries and in establishing TVNZ as an SOE in December 1988, provided an unusual opportunity to study the processes of strategic, organisational, and cultural change (and their performance outcomes) in a large broadcasting company.

Our study of the remaking of TVNZ was part of a wider analysis of the restructuring of government enterprises in New Zealand which followed the passage of the State-Owned Enterprises Act 1986. The story of TVNZ's transformation, while interesting and important because of the economic, political, and cultural significance of the broadcasting and communications industries, was by no means unique. Rather, it followed a general pattern similar to that experienced by other SOEs.[2]

Research Approach

To collect material for this book we conducted extensive interviews with many people who were intimately involved with the process of restructuring the broadcasting industry, the splitting apart and dissolution of the BCNZ into separate businesses and the establishment of TVNZ as an SOE. They included government officials, appointed members of steering and advisory committees on broadcasting, the last BCNZ board and TVNZ Ltd's permanent board; managers in TVNZ's television and other businesses; programme producers and schedulers; and knowledgeable outsiders. We also interviewed the chief executive and some board members of the Broadcasting Commission.

To conduct the study we were able to gain access to the files of the New Zealand Treasury and confidential company papers, including board papers and company plans relevant to the process of remaking TVNZ as an SOE. We also reviewed numerous public documents including commission reports, media accounts in newspapers and magazines, books and journal articles, internal newsletters, materials about the company and its programmes sent to advertisers, and published annual reports of the BCNZ, TVNZ Ltd, and the Broadcasting Commission.

Throughout several rounds of interviewing and analysis of documents we were careful to compare what we were being told in interviews with the story as it was contained in documents and other written sources. Wherever inconsistencies arose, or our understanding of particular events, decisions, or relationships appeared to be incomplete, we expanded our investigation to resolve the issue, at least in our own minds.

To ensure that our descriptions and interpretations of events were as reliable and complete as possible, the initial drafts of our manuscripts were read and commented on by the chairman of the board, senior managers of TVNZ, Treasury officials, and the executive director of the Broadcasting Commission. In some cases additional interviews took place to flesh out parts of the study.

The resulting comments were carefully considered and weighed when writing the final manuscript. Although a confidentiality agreement signed with TVNZ caused us to exclude commercially sensitive information, the agreement did not affect the analysis or materially affect the form or content of the discussion in the book.

Our approach is descriptive, analytical, and evaluative. We set out a careful description of events, process, and choices as well as an analysis of why certain choices were made. One of our objectives was to learn why things were done in the ways they were. In broadcasting, perhaps more so than in some other industries, there are many perspectives and viewpoints about particular events, processes, choices, and relationships. Wherever possible, therefore, we have tried to provide a balanced presentation of opposing viewpoints. We also present our own analysis and interpretations throughout.

Plan of the Book

To provide the necessary context and background we start in chapter 2 with a brief history of television broadcasting in New Zealand up to the formation of TVNZ as an SOE at the end of 1988. We set out a number of important milestones in the late 1980s which culminated in the decisions to restructure broadcasting in late 1988.

We then discuss a number of stages which were involved in the transformation of TVNZ into an SOE. The first stage is treated in chapter 3. This covers steps in the mid to late eighties to commercialise and revitalise the BCNZ, and focuses most attention on TVNZ as a division of the BCNZ. These steps were taken partly in response to global pressures on the broadcasting industry and partly in response to growing frustration with the rigid nature of the system in place at the time. It was during this period that a number of major reviews of the industry discussed in chapter 2 were conducted from which the Labour government's policy on broadcasting emerged. In chapter 4 we provide an overview of important sections of the State-Owned Enterprises Act 1986. We then examine the short, but crucial, establishment period from the end of August 1988 to the beginning of December 1988 as the BCNZ was prepared for its impending dissolution and TVNZ was prepared for its establishment as an SOE.

The third stage involves the shaping of the new SOE over the three-year period 1989–1992. Chapter 5 takes up the formulation and implementation of TVNZ's new competitive strategies as it readied itself for competition and went head to head against a new free-to-air competitor, TV3. The chapter also covers the formation of a number of strategic alliances TVNZ entered in broadcasting and related telecommunications. Chapter 6 describes and analyses how TVNZ's managers aligned TVNZ's organisational structures, incentives, and management processes with its strategies in order to secure a sustainable competitive advantage. The chapter also addresses the central issue of how TVNZ went about getting needed changes in its internal organisational culture.

Chapter 7 discusses a series of relationships within TVNZ and between TVNZ and outside parties. Relationships between the board and senior management, the board and shareholding ministers, and between the company and the new Broadcasting Commission are all examined. The chapter also provides an overview of the political, social, and cultural pressures which confronted the company and which involved programming mix and quality, editorial issues, and questions about privatisation and foreign ownership.

This is followed by an analysis in chapter 8 of TVNZ's financial performance for the four-year period from establishment as an SOE in December 1988 to the end of 1992. We also attempt to compare the financial performance of TVNZ before and after corporatisation, although this analysis is fraught with difficulty due to data limitations.

Our study concludes in chapter 9 with a detailed discussion of the management of change in the television industry, addressing in particular the significant institutional, organisational, and management factors contributing to the transformation of TVNZ from a state monopoly broadcaster into a successful electronic communications business.

CHAPTER TWO

A Brief History of Television in New Zealand

The Early Days

Although television was introduced in Australia in the early 1950s, it was not until 1960 that the New Zealand government, under pressure from the public, announced the introduction of a nationwide television service for New Zealanders. It was to be a state-owned monopoly operated by the existing New Zealand Broadcasting Service and funded by a combination of commercial advertising and a licence fee. The first broadcasts were made from Auckland in June 1960 for two hours a night, two nights a week.[1]

In 1962 the Broadcasting Act established the New Zealand Broadcasting Corporation (NZBC) as a state-owned (statutory) trading corporation to manage and develop public radio and television. A large development programme by the NZBC combined with individual initiatives to build and operate community television transmitters resulted in the rapid spread of transmission coverage and rapid increases in the number of television sets owned and in the size of the viewing audience.

The seventies ushered in a period of continued growth. Work started on the construction of a major television complex at Avalon just outside Wellington, and NZBC's transmission engineers succeeded in linking the whole country into one television network. Other major developments followed in quick succession. The existing black-and-white channel was converted to colour, a satellite earth station linking New Zealand to the rest of the world was opened, and applications were called for a warrant to operate a second national colour channel. Although the privately funded and operated Independent Television Corporation applied for a warrant, a newly elected Labour government made a political decision to maintain the state's broadcasting monopoly by awarding the second channel to NZBC. This decision to build a two-channel state-owned system was to have a considerable influence on the later development of television in New Zealand.[2]

A tumultuous period for broadcasting followed as the Labour government set about restructuring the entire broadcasting system by splitting the NZBC into a Broadcasting Council, two television corporations, a radio corporation, and an independent engineering body. In 1975 the NZBC's functions passed to Radio New Zealand; to Television One, which inherited the NZBC national network and the Avalon television complex; and to Television Two (South Pacific Television), which inherited little except staff. A brief period of intense competition between the two state-owned channels followed. By August 1976 Television Two had won 51 percent of the Auckland viewing audience.

In 1976, facing what it saw as expensive duplication of services in television news and current affairs and widespread viewer dissatisfaction with the services offered by the two channels, the new National government decided to merge all broadcasting services, including radio, into one corporation. The Broadcasting Act 1976 was passed, by which the Broadcasting Council was abolished, all state-owned broadcasting services were placed under the control of a new organisation called the Broadcasting Corporation of New Zealand (BCNZ), and a quasi-judicial Broadcasting Tribunal was established to regulate the industry. Some of the gains that had resulted from the earlier corporatisation were reversed by these events.

However, pressures to allow private interests to enter broadcasting persisted, and the National Party made a commitment to progressively increase this involvement in New Zealand television. Although sale of Television Two to private interests had often been threatened by the National government, it was not until the government actually instructed the Broadcasting Tribunal to call for applicants for a third channel that the prospect of a privately operated television broadcaster became a real possibility.[3]

Radical Change in the Eighties

The election of a reform-minded Labour government in 1984 ushered in a period of intense debate over the structure of broadcasting in New Zealand. Initially, reforms came slowly, but by the middle of 1988 the government had decided on a radically new broadcasting policy and set about its implementation. We turn now to several important milestones in the development of the government's new broadcasting policy.

Milestone 1: The Royal Commission of Inquiry into Broadcasting and Related Communications

In February 1985 the Labour government appointed a Royal Commission of Inquiry to inquire broadly into 'the institutions, operations, financing and control of New Zealand broadcasting and related communications and to report on what changes are necessary or desirable'. In September 1986 the Commission presented the government with a long rambling report in which it proposed expanding and tightening existing regulations.

The Commission recommended that cable television should be banned until all existing space on the VHF and UHF spectrum was utilised. Cable and direct broadcasting services should require licences from the Tribunal, programme schedules should be monitored for programme content, local content quotas should be introduced as a condition of holding a warrant, advertising should be restricted and the Tribunal should continue to have control over programme standards. The Commission also recommended many changes in the internal structure, operations, and management of the BCNZ and the types of services it provided.

The analysis which led to these heavily interventionist recommendations ignored many of the economic and technological pressures on the industry. It was also out of tune with the growing importance of neo-classical market economics that increasingly characterised the 1980s. As the Commission's recommendations were out of step with the government's programme of economic deregulation, they received little support from within government or from the industry itself. However, a minority report appended by Laurie Cameron, one of the commissioners, struck a responsive chord.[4] In his report Cameron argued that any analysis of broadcasting policy must recognise advertising and marketing by broadcasters as an important commercial activity, the effects of changing technology on competition, and the

9

organisational and management problems faced by the BCNZ. While the broad recommendations of the Commission's majority report were largely ignored and quickly forgotten, a number of the suggestions made in Cameron's minority report were later to be incorporated into government policy.

Milestone 2: Setting New Principles for Broadcasting Policy
By 1987 the government had started the process of deregulating the related telecommunications sector and was becoming increasingly frustrated with the inflexibility of the quasi-judicial Broadcasting Tribunal structure and the time it had taken to conduct hearings for the third television channel. A major inhibiting factor was section 80 of the Broadcasting Act 1976 which, among other things, required the Tribunal to consider whether the proposed service was in the public interest and the economic effects that new entrants would have on existing broadcasters.

The government's frustration caused it to turn its attention to the structure of the broadcasting industry. Following the re-election of a Labour government in 1987 and the appointment of Richard Prebble as Minister for State-Owned Enterprises, a joint examination of broadcasting options by officials of the Department of Trade and Industry and the Treasury began towards the end of the year. In April 1988 Cabinet made a policy decision, based on this work, to deregulate the broadcasting sector within a framework which would ensure that the government's social objectives in broadcasting continued to be met. Cabinet's expressed objective was to create an environment in which broadcasters would be able to compete by introducing new technologies and services in response to consumer demand.

To deregulate broadcasting Cabinet agreed that in principle it would

- remove legislative barriers to market entry involving economic licensing and the warrant system administered by the Broadcasting Tribunal;
- eliminate any special restrictions on the introduction of particular technologies (cable, satellite, and other transmission technologies were to be permitted subject only to general laws, spectrum regulation, and international obligations where applicable); and
- develop a method of spectrum allocation.

To meet its social objectives in broadcasting, Cabinet also agreed that in principle it would

- retain public ownership in broadcasting companies;
- ensure that foreign ownership and cross-media ownership was the subject of special restrictions;
- establish an independent broadcasting commission which would promote universal access, minority programming and programmes which promote New Zealand identity and culture by managing a system of competitive broadcasting grants funded by the public broadcasting (licence) fee; and
- set up an independent standards authority to promote and maintain acceptable standards of decency and behaviour on the part of the broadcasting media.[5]

This approach to reforming broadcasting was notably different from that proposed by the Royal Commission only a year or so earlier. Cabinet's reform principles promised to take broadcasting in New Zealand in a profoundly new direction.

Milestone 3: Implementing the New Policy Principles
Two committees were set up to advise the Minister of Broadcasting on how best to implement the decisions Cabinet had made in April. The Steering Committee on Broadcasting (Rennie Committee) chaired by Hugh Rennie, a Wellington media lawyer who was chairman of the BCNZ from 1984 to 1988, was charged with recommending an optimal organisation and financial structure for the BCNZ based on state-owned enterprise principles. An Officials' Coordinating Committee on Broadcasting (Officials' Committee), chaired by Jim Stevenson of the Department of Trade and Industry, was to deal with the details of how broadcasting was to be deregulated to create a competitively neutral environment and how the government was to achieve its social objectives in such a context. Two officials, one from Trade and Industry and one from Treasury, were made members of both committees to facilitate co-ordination. Although the work of the Rennie Committee preceded the work of the Officials' Committee, the combined recommendations of these two committees proposed a significantly different structure for broadcasting in New Zealand.

The unanimous recommendations of the Rennie Committee and the Officials' Committee were reported to ministers in early August 1988, accepted by Cabinet shortly thereafter and announced as government broadcasting policy. In a speech to the Advertising Institute of Wellington on 30 August 1988 Richard Prebble commented:

The Government is releasing both the Rennie and the Stevenson reports today. They are among the best official reports I have ever read on any subject and I would like to record the Government's appreciation of all those involved in compiling them.

In essence the reports indicate that the principles that the Government announced four months ago do provide a practical framework for a robust, deregulated, competitive, competent broadcasting industry. The Cabinet has determined to adopt both reports.

These committees made a number of recommendations which were agreed to by Cabinet on 22 August 1988. The most important were:

• *The BCNZ was to be replaced with two SOEs, Radio New Zealand Limited (RNZ Ltd) and Television New Zealand Limited (TVNZ Ltd). Cabinet determined that establishment, staff reorganisation, and transfer of assets was to be completed by 1 December 1988.*

An important aspect of the Rennie Committee's work was the careful consideration it gave to the organisational structure of broadcasting and, in particular, to the question of whether the BCNZ should be transformed into one SOE or two.

The Rennie Committee faced opposing arguments. The managers of TVNZ and RNZ, the Committee's consultants (Booz Allen and Hamilton Ltd), and officials favoured the split, but the BCNZ's board, chief executive, and director-general of television argued against separation.[6] However, after considering a range of arguments involving (1) possible synergies and economies of scale between television and radio; (2) potential conflicts of management style, business operations, image, and market position; (3) competition for advertising revenues; and (4) problems of cross-media ownership and internal cross-subsidisation, the Committee concluded that 'more would be gained than lost through separation of the TVNZ and RNZ businesses' and recommended that 'two separate businesses be created with independent boards'. Their report (p. 63) concludes:

> The needs [of the two businesses] are different and both TVNZ and RNZ would gain from the specialist focus of management and boards on their target industry namely on television and radio. This was seen by the Committee as the most compelling argument in favour of the separation and although it is not easily quantified it is nevertheless a key to the success of TVNZ and RNZ in the future environment of competition, falling relative advertising revenue and rising investment demands to keep up with the latest television and radio technologies.

The Rennie Committee also recommended against establishing TV1 and TV2 as separate SOEs. In the Committee's opinion separation would simply result in the loss of substantial business benefits arising from the joint use of resources, the ability to programme the two channels in a complementary manner, and the attractiveness to advertisers of a two-channel network controlling a larger proportion of airtime.

• *BCNZ's downstream transmission and engineering assets (including distribution links) were to be appropriately divided between TVNZ Ltd and RNZ Ltd and operated as separate businesses and cost centres within the respective SOEs. Television transmission sites and facilities were to be vested in a separate subsidiary company and profit centre to promote transparency of operations and arm's-length transactions with broadcasters.*

TVNZ's management argued vigorously that downstream signal distribution links and broadcast transmission facilities were strategic assets that should remain under the ownership and control of TVNZ. They believed that separation of the assets of the Broadcasting Services Division of the BCNZ from TVNZ Ltd would make the new television company vulnerable to hold-up by whoever controlled transmission and (most importantly) would stop it capitalising on significant opportunities brought by the deregulation of the telecommunications sector such as the use of the distribution network to create the backbone of a general-purpose telecommunications system competing with Telecom and other carriers. The Rennie Committee was convinced by these arguments and recommended that the control of signal distribution and transmission assets remain with TVNZ. In addition, if the control of these activities did not pass to TVNZ Ltd, there was no assurance that it would continue to purchase services from an independent distribution and transmission company. Indeed as Hugh Rennie has since pointed out:

> . . . there was every reason to think that TVNZ for strategic reasons, would not purchase such services and would elect to provide these services itself. The destruction of value in the existing business which this would create was not acceptable to the restructuring committee.

For similar reasons the committee recommended against separation of TVNZ's upstream programme production facilities. The committee argued that a forced divestment would result in an erosion of New Zealand's production base just when the demand for English language television programming was increasing. Because TVNZ was a low-cost television producer by world standards, the new company would be well placed to capitalise on these opportunities. The committee put the argument in their report as follows:

> . . . the total separation of production from TVNZ [may] irreparably damage an industry which may be just a few years away from commercial prosperity as an export-orientated production base. If this view is correct, the value maximising strategy would be to retain adequate production facilities within TVNZ, as well as commissioning enough internally produced programming to keep them thriving while the export market is developed and then benefiting from the external sales revenues which are generated. This benefit can be realised either as a continuing profit stream or through the sale of production units once they are no longer dependent on internally commissioned work.

The Officials' Committee generally agreed with these recommendations but recognised that continued ownership and control of key transmission sites and assets and production facilities would provide TVNZ with certain competitive advantages. However, to constrain anti-competitive use of the transmission assets in the future, officials made a series of recommendations. First, all TVNZ television transmission sites and facilities should be vested in a separate subsidiary company. Second, the transmission subsidiary should be required to develop clear contractual terms and pricing structures for access to its network. Third, the new board of TVNZ Ltd should be asked to affirm the neutrality of its subsidiary in dealings with competitors.

Officials believed that this set of measures, when added to the general protection available to competitors in section 36 of the Commerce Act 1986 (which forbids a firm holding a dominant market position from using that position for the purpose of preventing or deterring entry by competitors) would allow TVNZ Ltd to capitalise on new uses for its transmission assets while at the same time constraining anti-competitive behaviour with respect to other broadcasters.

• *The government's stated public service broadcasting objectives were to be implemented primarily through the establishment of a broadcasting commission.*

The Officials' Committee recommended that the commission be established as an independent body corporate to administer a system of contract funding to promote programming which would reflect and develop New Zealand's identity and culture; to extend coverage of broadcast signals to remote communities; and to ensure that minority interests were catered for. The commission was to be funded primarily by a public broadcasting fee (expected to be around $55m in 1988–89) and fixed by regulation under the Broadcasting Act.

Officials also recommended that some activities such as the New Zealand Symphony Orchestra, parliamentary broadcasts, the international shortwave service, and educational broadcasts should be funded from general taxation, thereby increasing the amount available for funding New Zealand content in television. If all these activities were to be funded from the public broadcasting fee then of the $55m expected, officials estimated that approximately only $11m would remain to fund local content in television or other commission priorities. This was too small an amount to be effective.

The Officials' Committee favoured funding the government's commitment to public service objectives through a system of competitive contracts. The alternative of direct regulation of schedule content was rejected because officials believed that it would result in higher costs on broadcasters and consumers, limit competition by making it more costly to enter broadcasting markets, be costly to regulate and administer, retain inefficient cross-subsidies within broadcasting organisations, and interfere with the ability of New Zealand broadcasters to compete head to head with foreign broadcasters beaming signals directly into New Zealand via satellite in the future. Rather, officials argued that a contracting approach offered the advantages of 'competitive neutrality, transparency, targeting, potentially lower cost and responsiveness to consumers'.

• *The new broadcasting SOE's and TVNZ Ltd in particular, would have an obligation to encourage New Zealand-made programmes that reflect and develop New Zealand's identity and culture.*

To this end the Officials' report recommended that conditions be placed in TVNZ Ltd's Statement of Corporate Intent (SCI) to ensure that the company had this as its *predominant* objective. In the interests of competitive neutrality, officials recommended that conditions in TV3's warrant from the Broadcasting Tribunal concerning local programmes and Maori content should continue to apply until 31 March 1992.

It is interesting to note that, in order to meet the government's public broadcasting objectives, officials were not content to rely either on commission funding of local content or on the commercial incentives TVNZ Ltd would have to differentiate and brand their channels through local content programming. Jim Stevenson explains why the Officials' Committee recommended that a social objective be included in TVNZ Ltd's SCI:

> There was a view that if the Crown's reason for continued ownership of broadcasting assets was to keep alive the social objective then that should be reflected in the objectives of the company. Otherwise it should be sold—not that the Labour government ever had any intention of selling it. In that way it was seen as a moderating influence on what might be purely commercial rate of return or dividend objectives. If TVNZ went for a greater proportion of more expensive local programming over cheaper imported programming they couldn't be expected to reach the norm of what public companies earn.

Because this recommendation contemplates a level of local programming over and above that which would be undertaken on purely commercial grounds, it seems out of character with the other market-oriented proposals made by the Officials' Committee and the Rennie Committee. One argument presented by an involved official was that its inclusion reflected the need to place a 'saleable package before ministers'. Even so it was to become a contentious issue which resulted in the board of TVNZ Ltd being caught in a cross-fire between the Department of Trade and Industry and the Treasury. We will return to a further discussion of this issue in chapter 7.

• *High behavioural standards by broadcasters were to be promoted and maintained.*

Officials recommended that broadcasters continue to be required, under the Broadcasting Act, to maintain acceptable programmes and advertising standards and that an independent Broadcasting Standards Authority should be established to promote standards, hear complaints, and impose sanctions.

• *The level of foreign ownership in respect of radio and television firms was fixed at 15 percent but the Minister of Broadcasting would have discretion to increase the maximum to 25 percent for radio broadcasters.*

Under the old broadcasting regulations overseas ownership was restricted to 5 percent, although the Broadcasting Tribunal had the power to give special approval to increase that level to 15 percent. The main perceived benefit of these restrictions on broadcasters was to protect and enhance New Zealand identity and culture. The costs to broadcasters and consumers resulting from these restrictions included a potentially higher cost of capital, an inability to exploit fully economies of scale and the potential for competition in the broadcasting market being prevented because of an inability of overseas investors to enter the market.

Weighing its assessment of these costs against the perceived benefits, the Officials' Committee recommended that overseas investment restrictions for the broadcasting industry be fixed at 25 percent (with the exclusion of narrowcasting activities where standard foreign ownership provisions should apply). This would bring broadcasting restrictions generally into line with Overseas Investment Commission procedures under the Overseas Investment Act which required ownership by overseas interests in excess of 25 percent to have Commission approval.

However, this recommendation did not go forward and Cabinet instead approved a cut off at 15 percent, the level to which foreign ownership could be approved by the old Broadcasting Tribunal. As will be discussed in chapter 7, the National government which succeeded the Labour government was later to remove all foreign ownership restrictions in 1990.

Conclusion

Television broadcasting has gone through many changes since it was first introduced in New Zealand. The most radical changes of all were proposed in the late 1980s as part of the Labour government's wider efforts to restructure the commercial activities of government enterprises. In broadcasting two government-appointed committees made many far reaching recommendations which would completely overhaul the structure of the industry in New Zealand.

The Rennie Committee focused its attention on recommending organisational arrangements which would maximise the value of the broadcasting assets owned by the state. Through a broad-scale analysis of the strategic options facing the new SOEs and in particular TVNZ, it was able to leave the new TVNZ Ltd with the flexibility and asset base necessary to deal with competitive threats and to pursue aggressively opportunities in both domestic and international markets. The Officials' Committee, on the other hand, concentrated on ensuring that the new broadcasting SOEs would have to compete on an equal footing with private broadcasters in a competitively neutral legislative environment. The establishment, deliberations, and reports of these two committees provide an important example of how the process of determining the competitive environment and the organisational framework for the setting up of state-owned enterprises can be successfully managed. Much of the credit for the structure New Zealand now has in broadcasting resulted from the implementation of the work of these two committees.

Achieving the recommended changes required the passage of three pieces of legislation. First, the Broadcasting Amendment (No. 2) Act 1988 (the 'Restructuring' Act) provided for the dissolution of the BCNZ and the transfer of its assets, contracts, and liabilities to the new broadcasting SOEs. Second, the State-Owned Enterprises Amendment (No. 4) Act 1988 attached TVNZ Ltd and RNZ Ltd to the Act's schedule of state-owned enterprises and added a section to protect the editorial independence of broadcasting SOEs. Third, a new Broadcasting Act 1989 provided for the establishment of the Broadcasting Commission and the Broadcasting Standards Authority. It also effectively deregulated the broadcasting industry.

Both committees recognised that acceptance and implementation of their recommendations would place the Crown's television broadcasting assets at risk. As the Rennie Committee's report put it:

> The structure and changes proposed for TVNZ implies a high risk strategy for the Government to manage. The preference for the Government would normally be to adopt a low risk strategy of retrenchment which might force TVNZ to resist the competitive challenge and adopt an insular rather than an export-oriented approach. Indeed one could envisage a potential tension between the shareholder and the TVNZ management with the former trying to constrain TVNZ's high risk and overseas investments. Flexibility and commercial disciplines should be adopted at every level from the shareholder to the programme editor.

Of course in most respects television was already at risk from the approval of the third television channel, the ongoing deregulation of the related telecommunications industry, and rapid technological change. Perhaps the greater risk, however, would have been to adopt a defensive strategy and retain the status quo.

We pick up the discussion of the implementation of the Labour government's policies in broadcasting in chapter 4. In the following chapter we discuss changes made in television broadcasting in the BCNZ prior to corporatisation.

CHAPTER THREE

Commercialisation Prior to Corporatisation

In the period before corporatisation a number of improvements were made in the commercial management and operation of TVNZ as a division of the BCNZ. In this chapter we consider the progressive commercialisation that took place in the BCNZ and TVNZ, focusing primarily on the period 1984–1988.

Problems with the Broadcasting Corporation of New Zealand

The Broadcasting Act of 1976 established the BCNZ as a statutory corporation of government with a monopoly over television broadcasting in New Zealand. Its operations were financed by revenues from advertising and from a public broadcasting (licence) fee set by the government and levied on the owners of television sets.

Section 3 of the Broadcasting Act of 1976 set out the following objectives to be pursued by public broadcasting and broadcasters:

(a) To maintain and develop broadcasting as a system of human communications, to serve the people of New Zealand:
(b) To obtain, produce, commission and broadcast a range of programmes which will inform, educate and entertain:
(c) To ensure that programmes reflect and develop New Zealand's identity and culture; and that programmes are produced and presented with due regard to the needs of good taste, balance, accuracy and impartiality and the privacy of individuals.

Under the Act the BCNZ had mixed objectives. It was expected to operate in a commercial manner while promoting public broadcasting objectives through the operation of its television and radio stations. However, it was the BCNZ that decided how to pursue public service broadcasting objectives and to what extent it would cross-subsidise local programme production from its commercial revenues.

The combination of commercial and social objectives created a number of problems. First, it made it extremely difficult for the government to effectively monitor BCNZ's performance and to determine if it was successfully meeting either its commercial or social objectives. Second, the corporation's mixed objectives created a dilemma for board members and managers who had to make investment, expenditure, and borrowing decisions. Board members, who considered themselves as trustees of the public interest rather than as the directors of a commercial company, generally dealt with this dilemma by down-playing the weight they gave to commercial performance in making decisions.

The selection of BCNZ board members for reasons of geography, gender, cultural or literary background and politics more than business experience resulted in boards with diverse interests without a strong commitment to commercial efficiency. A production-oriented culture developed. By the middle eighties, the Rennie Committee noted that approximately three-quarters of all TVNZ staff were employed in programme production and more than half of its assets were directly allocated to this activity. The Rennie Committee further observed that significant under-utilisation and inefficient use of resources resulted from this in-house concentration of programme production.

To make the matter of performance monitoring and accountability even more confused, under the Act the BCNZ was also required to provide policy advice to the Minister of Broadcasting on broadcasting matters. As it was also the dominant player in the broadcasting industry this placed the BCNZ in a serious conflict-of-interest situation. It was required to act as an adviser to government on broadcasting policy development and to be a sensitive public broadcaster while, at the same time, it was expected to operate commercially in radio and television markets. With increasing competition in radio and the prospect of competition in television, the BCNZ's ability to juggle and reconcile its conflicting roles became increasingly problematic.

In any event, the BCNZ's freedom to manage its operations along either public interest or commercial lines was inhibited by the structural and operational strait-jacket imposed by the Act. First, the original Act prescribed an unwieldy top-management structure on the corporation. It provided for two directors-general of television (one for each channel) and a director-general of radio but did not allow for the appointment of a chief executive above them. Rather than amending the Act an attempt was made to work around this organisational problem by appointing an executive chairman who functioned as both chairman and chief executive. This proved to be a singularly unsuccessful arrangement and in 1985 the Act was amended to make changes to the structure of top management.

Second, the Act gave overlapping and conflicting powers to two different ministers. The approval of the Minister of Finance was required for all major borrowing, expenditure, and investment decisions. However, the Minister of Broadcasting had the power to issue policy directives which had operational, financial, and managerial implications. As a result, it was sometimes difficult for the board of the BCNZ to understand to whom it was accountable, although as a practical matter accountability on most issues was to the portfolio minister, the Minister of Broadcasting. The Minister of Finance tended to be consulted infrequently and only when strictly necessary.

Third, the Act made the BCNZ subject to the control of the Broadcasting Tribunal which regulated entry into broadcasting, ownership, area coverage, and programming. Through a system of broadcasting warrants the Tribunal imposed detailed operational restrictions on the levels of advertising, the hours radio and television stations could operate, and the level of local content to be broadcast. This system severely inhibited the operational and commercial flexibility of all broadcasters, including the BCNZ. An attempt to establish a new legal subsidiary to undertake drama production, South Pacific Pictures, in the last days of the BCNZ illustrates the extraordinary difficulties involved in obtaining approvals for commercial reorganisations of this kind. For months it proved impossible to obtain the necessary ministerial consents because of Treasury opposition arising from concerns about competitive neutrality and the structure of the BCNZ, which officials argued was not well suited for the effective delegation of authority with commensurate responsibility. A Treasury official indicated that one reason for the delay was that this application gave the Minister of Finance a rare opportunity to look at the overall performance of the BCNZ. South Pacific Pictures started a year later than planned because of these difficulties. The chairman and chief executive of the BCNZ and the director-general of television also committed substantial efforts to obtaining a simple approval for this business activity. About this episode the chairman of the BCNZ, Hugh Rennie has since remarked: 'In general, it was a lot simpler not to try such business initiatives, than to commit the time and resources trying to obtain the necessary approvals.'

Lastly, as a broadcasting monopoly TVNZ had capitalised on its central position between advertisers and viewers and its lock on television broadcasting. However, with deregulation and a new entrant, TVNZ would have to compete for spectrum and for revenue. On the other hand new opportunities would result from the deregulation of telecommunications. This meant that BCNZ would be able to utilise its transmission network for purposes other than transmitting broadcasting signals, provided that it was not precluded from entering broader electronic telecommunications markets.

A Treasury official who was actively involved in the later restructuring of broadcasting summed it up as follows:

> We knew there were problems with the BCNZ structure. There were clearly conflicting objectives for the organization. It was the collector of the public broadcasting fee, the advisor to the government on broadcasting matters as well as being a player in the commercial industry. At that stage they were a monopoly provider of television service. There was a concern about the public broadcasting fee going into the organization. You couldn't actually isolate what public broadcasting services were being bought with these funds. The only commercial overview that took place was when an application for an increase in the broadcasting fee was made or the corporation wished to borrow money.

Beginning the Process of Organisational Change

After the election of the Labour government in 1984 changes started to be made at the BCNZ to improve the commercial management of the corporation. These changes were triggered by two actions taken by the government. The first was to appoint a new chairman and to replace almost completely the BCNZ board. Hugh Rennie was appointed chairman in late 1984. New board members were appointed by the government in 1984 and 1985 and Brian Corban was appointed as deputy chairman in May 1985.[1] The chairman, deputy chairman, and other board members found a corporation they considered to be in poor shape to respond to future competition, the rapid changes taking place in broadcasting technology and the nature and shape of the international television industry. Hugh Rennie comments:

> The BCNZ was hopelessly overstaffed and there were no internal financial controls in my view. I went from a small and very lean and very efficient newspaper company to this extraordinary organization which seemed to move forward of its own momentum because in television, which was its main operating area, it was operating in a monopoly situation. It could sort of muddle along, it didn't have to look too hard at where it was. To turn this thing around would involve bringing in financial controls and major changes in staffing, culture, attitudes, and objectives.

Some senior managers felt that commercialism would damage the corporation. But there was one matter on which the board and management were united—all had a commitment to persuading the government to allow a substantial increase in the public licence fee. The continued development of public service television, the ability to cover rising costs in the Symphony Orchestra, the introduction of new radio services (including plans for Maori radio), and an extension of signal coverage to uneconomic areas were all viewed as contingent on raising additional revenues from licence fees.

Rennie and the BCNZ board believed that TVNZ was an organisation driven by the desires of producers to make television programmes rather than by audience demands or commercial realities. Rennie felt that the management of TVNZ had little idea about what the future held in store for television. The fact that there was no real head office for TVNZ was, in Rennie's view, symptomatic of an organisation seriously at risk of being swamped by waves of new technology and competitors surging into broadcasting. Even though the BCNZ had been operating as a statutory corporation for eight years, BCNZ managers and staff still had a government department mentality. Rennie explained:

> In 1984/85 the BCNZ was still in conceptual terms a government department and its financial thinking was certainly not private sector based. The corporation had something like $25m of cash sitting in the bank and there were no borrowings. Superficially, this appeared to be a very strong position until I discovered that it had been accumulated by failing to tackle key equipment and facilities issues. Auckland television operated out of thirteen buildings some of which were slums. From memory, it was 1986 before I saw a cash flow statement. The $25m in cash had run up to something like $33m before we could really begin

24

to turn it around. It was best described as a piggy bank mentality. It was explained to me that one day the corporation's trading would deteriorate and then we'd be glad we had $25m in the bank.

Because of unreliable or non-existent accounting, planning, and reporting systems it was very difficult for the more commercial members of the BCNZ board to get to grips with the problems of the organisation. Brian Corban, the deputy chairman of the BCNZ, described his reaction after joining its board:

> I gradually became frustrated because of the mishmash of business, bureaucratic, cultural, and political issues that came to the board. The whole culture of the BCNZ, the nature of the executives and their backgrounds, was basically bureaucratic and only semi-commercial. The financial reports to the board seemed always undimensioned, always incomplete. A capital expenditure [request] would appear out of the blue rather than out of a strategic plan. You couldn't test the expenditure because there was nothing to test it against apart from a bevy of six executives who appeared before the board saying 'We must have this'. I found that frustrating. There was a lack of context, perspective, and dimensioning that arose from a lack of structure.

These views of the organisation were shared by Laurie Cameron, one of the four commissioners of the Royal Commission of Inquiry into Broadcasting which sat during 1985 and 1986. In his minority report Cameron wrote:

> It was clear from the Commission's research that the BCNZ has remained, in both form and attitudes, a Government Department for too long and its staff has continued to be trained and promoted accordingly. So long as the BCNZ remained a monopoly in television and a near monopoly in radio and despite adverse financial results during the 1970s, the motivation was lacking to make the drastic changes required to meet the increasingly competitive environment for radio and the increasing costs of local New Zealand television production.

The second action taken by the government involved amendment of the Broadcasting Act to reorganise the top management structure of the BCNZ. Based on a recommendation of the newly appointed chairman, the positions of chairman and chief executive, which had been combined and held by one person, were separated, and an executive committee of the chief executive, the directors-general of television and radio, and the general manager of resource services was established. Indicative of the inflexibility of the statutory corporation form of the BCNZ was the fact that even this relatively simple reorganisation of the management structure required legislative amendment.

To gain a measure of commercial flexibility the board also lobbied the Minister of Broadcasting to reduce the powers of the Broadcasting Tribunal. At that time the Tribunal regulated the hours broadcasters went to air and the levels of advertising. This power caused the BCNZ, and particularly TVNZ, extraordinary difficulties. The board asked the Minister of Broadcasting to amend the Broadcasting Act to reduce the Tribunal's power to control the number of hours on air and when advertising could be carried. Even though these changes were resisted by the Tribunal, the Act was amended to give the corporation increased commercial flexibility.

Second, because of the upcoming retirement of a number of top executives, the board was provided with an opportunity to bring in new management. In particular, both the chief executive, Ian Cross, and the director-general of television, Alan Martin, were due to retire. In early 1986, following a worldwide search, the board appointed Julian Mounter, an Englishman, as director-general of television and a little later Nigel Dick, an Australian, as chief executive of the BCNZ. With the Royal Commission inquiry well under way and the third television channel hearings proceeding, the BCNZ desperately needed experienced media executives who were familiar with the realities of commercial television and understood business competition and planning. As Hugh Rennie put it:

> One of the tragedies of the BCNZ was that no attempt had been made to expose individual executives to this [business competition and planning]. There were serious deficiencies in training for succession and in identifying and developing executives with appropriate business skills. Worst of all, there was no real experience or understanding of the skills involved in running a commercial media business.

Indeed, it was only after the arrival of these two men that the Royal Commission, according to Laurie Cameron, began to

> receive a more realistic picture of the BCNZ and its organisation and at that time was given some vision of the future for broadcasting in New Zealand as a result of advances in technology and the increase in future competition.

However, these two new top executives did not hit it off together and they were soon at loggerheads over broadcasting policy, television operations, and the extent to which the corporate office should be involved in the day-to-day operations of the television division. Drawing on his extensive experience as a senior executive in the Australian television industry, where the industry had managed to resist change because of the considerable power of its media giants and their close relationship with government, Nigel Dick started a campaign to improve the BCNZ's relationships with politicians in order to defend the BCNZ dominance. This involved trying to stop the third channel (or at least ensure that it was set up in such a way that it did the least injury to the trading position of the BCNZ), to gain a larger public broadcasting licence fee to maintain programme services and standards in the face of significant revenue downturns he believed were on the horizon, and to resist what he considered to be excessive commercialisation of television in New Zealand.[2]

Julian Mounter on the other hand wanted to make more fundamental changes to television broadcasting. He believed that to survive TVNZ had to lead change through strategic thinking, planning, and management. He argued that attempts to resist change or emulate the BBC model of television broadcasting were a recipe for disaster. Moving into television after working as a newspaper journalist, Mounter had gained experience in the United Kingdom working for regional television, the BBC, ITV, Thames TV, and Thorn-EMI as producer, director (programmes), manager, and as company director of an independent production company. At Thorn-EMI, Mounter helped set up a new satellite and cable programmes division. As director of programmes and production for the division he was responsible for designing and managing the 'on-air' look, staffing and programme production and budgeting systems for three programme channels. The knowledge gained in these positions gave Mounter a global perspective on the threats and opportunities that were facing television broadcasting around the world.

A series of increasingly bitter battles inside and outside the boardroom ensued, triggered by what the director-general of television saw as foot-dragging and resistance from the BCNZ's central bureaucracy for his plans for change and by what he considered to be the corporate chief executive's unwarranted interference in the day-to-day operations of the television division. An important flashpoint was Mounter's decision to remove Phillip Sherry as the anchor of the evening news as one part of an attempt to change the image of TVNZ. Nigel Dick, however, was strongly opposed. He believed that Sherry was respected and even beloved by New Zealanders and that his removal would significantly damage ratings. In the end, on this, as on a number of other issues, Mounter gained the support of the BCNZ board for his views. In the face of what he considered to be a diminution of his authority with respect to the television division, the chief executive retired from the BCNZ, his contract was paid out, and he returned to Australia. Of these two men, therefore, it was Mounter who ultimately was to have the greatest influence on the direction of TVNZ and television broadcasting in New Zealand.

Mounter reached basically the same judgement as Hugh Rennie and Nigel Dick about the poor shape TVNZ was in to deal with competition. His assessment was that even though TVNZ had a number of very competent, professional broadcasters, it was ill prepared to meet the challenges of competition or to cope with the technological changes which were sweeping through the industry worldwide. He describes the organisation he found:

> There was vast over manning and far too many facilities. Many were sitting open and unused for large chunks of time. This was something I had battled against in British television and I was appalled to find it worse here. I knew there were massive savings to be made. Walking into the Shortland Street studios in Auckland was like walking into a television museum. There were cameras there that I had only seen in the Victoria and Albert Museum in London. TVNZ in Auckland was spread over 13 or 14 buildings, eight to ten of which had canteens. Cars and minivans were running between buildings. In logistical terms it was awful. Everybody knew there was likely to be a third channel. Somebody had to do something to get this place in shape. I thought we were going to be wiped out.

Compounding the problem were inadequate and ineffective financial systems, poor internal disciplines and restrictive management practices. Mounter recalls:

> When I first sat down at the director-general's desk there were these great piles of paper. For example, every trip to Wellington or Auckland or overseas had to be signed off individually, any hiring of a part-time person on a drama had to be signed off. TVNZ practised a policeman style of management. In the end you have to have delegation of authority because until producers care about money as well as creativity, they can always out-think the policeman. I also ran into simple outright defiance

in some areas. The answer to this anywhere else would have been 'OK, if you don't want to work for me, goodbye'. But you couldn't fire anyone and you couldn't hire anyone. There was this committee that would consider internal appeals [against appointments] and they could say 'I'm sorry, we want this person because we think he or she is more qualified'.

Mounter recognised that in order for any large-scale reorganisation of TVNZ to be sustainable, the structure of television broadcasting in New Zealand would have to be changed. While he made this his first priority he was still able, with support from Hugh Rennie and a number of BCNZ board members, to drive through a number of important changes to prepare TVNZ for future competition from a third channel. However, change was met with significant resistance and antagonism from management. Mounter attributes this to the fact that he was a 'Pom' and an unknown quantity to managers; that TVNZ's managers had little outside commercial experience; and that a lot of them did not want to see the present system of promotion and progression in the organisation disrupted.

The first real challenge, therefore, was to deal with resistance to change. Mounter tackled this problem by bringing in new managers from outside. His objective was to put in senior managers who would gain the support of middle managers who were frustrated by the present system. Mounter explained:

> In the first year and a bit, I changed over fifty senior management jobs. A lot went out of the organization. I gave them plenty of time, but they resisted, they argued, they wouldn't have it. I knew from my experience in Wales that it would have been disastrous to hire [too many] foreigners. I tried to hire Kiwis by bringing Kiwis back from abroad with commercial experience. I brought something like twenty Kiwis, people like Harold Anderson and John Barningham, back into senior management positions. And I hired Kiwis here and I promoted from within. This allowed us to rapidly build a team that agreed that there had to be a change.

A second challenge was to quickly upgrade TVNZ's technology and to physically modernise the television operations. On this Mounter had the support of the board and the new chief executive, Nigel Dick. As Hugh Rennie was later to write:

> Many BCNZ assets were inappropriate in design, location or operating costs to compete against new broadcasters using modern technology. Some were a woeful legacy of earlier political and financial compromises,

which had deferred replacement of essential equipment and left the Corporation over-staffed and in some areas under-skilled. Others resulted from equally woeful excesses in the BCNZ spending on unnecessary or uneconomic assets.[3]

An Auckland Television Centre estimated to cost $90m, which included $25m for new equipment, was on the drawing boards. Land for a centre had been purchased in 1985 with the assistance of the deputy chairman of the BCNZ, Brian Corban. Mounter believed an Auckland centre was needed urgently in order to pull the network's organisation together into a coherent whole prior to the introduction of competition from a private third channel. Because of the regional make-up of the board there was initially some resistance to this large expenditure in Auckland. Although the chief executive of the BCNZ, Nigel Dick, supported the concept of a new building in Auckland, he did not support its proposed size and dimensions which he saw as overly expensive and grandiose. Some board members argued that if funds were to be spent upgrading Auckland facilities then funds should also be spent on the Christchurch and Dunedin stations.

However, board support was finally won, and the headquarters of TVNZ were moved to Auckland. Although this was an unpopular decision among Wellington television staff, Mounter argued that it was important for TVNZ's headquarters to be located in the commercial centre of New Zealand, in the middle of the largest mass of the country's population and where trends in lifestyles and audience demands could be most easily ascertained. It would also be in Auckland where the looming battle with TV3 would be fought. An added bonus was that it was well away from the politicians in Wellington. The move to Auckland was a critical step in the development of TVNZ's new corporate image and culture.

To speed up construction, Mounter appointed Graeme Wilson, then deputy director of news, as project manager. He reasoned that a news man who would have to live in the building would have an incentive to push it through quickly. (The new centre was subsequently brought in on time and on budget with the bulk of staff moving into the new building by September 1989.) In addition to providing a new headquarters, the centre was set up to originate news and current affairs programmes, to produce other programmes that can use the same resources, and to provide transmission and supporting facilities for the operation of Channel 2.

A third challenge was to capitalise on the fact that TVNZ controlled two television channels. Mounter described the two channels in 1986 as 'look-alikes' because both channels were being programmed in a similar way. In Mounter's view this gave away the significant competitive advantage TVNZ possessed in its control of two network channels. He argued that the audience choice and the market positioning of TVNZ would be enhanced if the channels were programmed and scheduled in a complementary fashion. This helped to set in motion a process of updating the programming and scheduling of TVNZ's two channels and developing a distinctive 'on-air' character for each.

With competition from a private television network just around the corner, the fourth challenge was to overcome complacency and a lack of entrepreneurship in the organisation and to inject an aggressive, competitive mind-set into managers and staff. Mounter commented:

> Managers couldn't see the need for this structural thinking change. I said 'OK, I want you to go overseas and buy the rights to important quiz show formats because if we don't they will'. They would answer 'We don't have to do that now, TV3 won't be on air for three years'. I'd reply 'No, we have to do it now'. That sort of aggression was lacking in the company. Nobody really wanted to be aggressive. At the same time competition was missing in the whole of New Zealand society; people thought of it as dirty pool.

It was in this way that the process of organisational change started in TVNZ prior to its establishment as an SOE. Mounter provided a critical stimulus by persuading a significant proportion of the people who worked at TVNZ that change was inevitable and was being driven by technological advances and global competition. He argued that it was futile to try to protect the monopoly legislatively in a world where direct broadcast delivery to homes from high-powered satellites was possible and where advances in digital technology were causing a revolution in electronic communication. What the organisation needed to do, in his opinion, was to develop a strategy for how it was going to succeed in a rapidly changing environment. Graeme Wilson, then general manager of TVNZ Networks, commented:

> Go for it was Julian's approach. He was absolutely determined that management effort of TVNZ wouldn't be wasted on bureaucratic or other efforts to resist change. That the whole of the efforts of management and

31

others in TVNZ ought to be directed towards achieving change and success in the changed television environment. That was very different to the way in which other changes in television I had experienced had happened where people in the organization plough furrows with their heels as they are dragged unwillingly into the new structure. When you can unleash that sort of force in an organization that has good creative people then you can really get things humming and things definitely hummed around here.

Learning to Think Strategically

Some thinking about how TVNZ would meet competition from a new private television network had been done prior to Mounter's arrival at TVNZ in 1986 and resulted in a document called *The 1986/ 87–1988/89 Strategic Plan*. Although this document provides a useful assessment of TVNZ's strengths and weaknesses at this time, it had a relatively narrow focus on how competition from the third channel could be met. The plan did not communicate a sense of real urgency, nor was it particularly broad in scope or innovative in its approach. The director of resources, Rod Cornelius, provided the following retrospective assessment:

> Des Monaghan [the director of programmes and production] and I were sitting down doing plans saying how the new channel would take a third of our business in the first year. We didn't have any strategy, we just knew it was going to hit us. And Julian Mounter came up with what to do. How to make the company work. How to sharpen us up. We were insular, we were looking inwards. He introduced a whole new way of thinking about television and got us thinking not only about TVNZ in competition in New Zealand but also about our wider future.

With his team of senior managers starting to take shape, Mounter led a series of senior management workshops starting in July 1987. After reviewing the external and internal environments of TVNZ and financial models which were based on BCNZ corporate planning models, it was agreed that TVNZ must do four things: first, it would strive to position its two television channels at number one and two in a three-channel market; second, it would work to counter the impact of further fragmentation of the audience by becoming involved in profitable opportunities for new broadcasting or narrowcasting services; third, it would search for new sources of revenue at the same time as it worked to maintain and develop the core business; and lastly, it would seek to reduce costs while maintaining quality. Julian Mounter saw the strategic issues facing TVNZ as follows:

> The strategic problems were the coming of TV3 and the fact that the market would fragment with more players, the possibility that we might be overtaken in delivery services such as satellites which would steal our market, and how we could preserve our local identity and culture. They were very clear to me before I arrived and they didn't change. Those are still the issues today.

With essential strategic directions set, a further series of management workshops on winning the audience, sales and marketing, new revenue opportunities, improving efficiency, and doing things differently were held over the following months. Each workshop involved a team of managers in the planning process and developed projects for further action. The process adopted in each case was similar. First, teams were asked to put themselves in the competition's shoes. For example, how would they programme TV3 to win the audience and push TVNZ out of the number one and two positions in the ratings? How would they organise the TV3's sales and marketing operation to win advertising revenue? How would they organise and manage TV3 for efficient and effective operations? The simulated TV3 strategies were then analysed to identify key factors requiring attention within TVNZ if they were to be countered. Within this context TVNZ's current operations were critically reviewed to identify important changes that needed to be made. In each case this approach created a 'fresh start' or a 'zero-base' from which innovative planning could proceed. The objective was to leave TV3 no 'easy marks'.

Awaiting More Radical Reform

By December 1987 the major elements of TVNZ's new strategy had been determined and changes started to be made as TVNZ readied itself to meet the challenge of competition. But TVNZ was still part of the BCNZ, which continued to operate under the restrictions of the Broadcasting Act 1976 and under the control of the centralised BCNZ bureaucracy. In this environment there was only so much that TVNZ, as a division of the BCNZ, could do to reorganise itself. As Brian Corban, who was deputy chairman of the BCNZ at the time, has since commented:

There was some reorganization, some adoption of more commerciality, some new energy, and some breaking of the bureaucratic mould, but all enormously, enormously far short of what had to be done. Change was frustrated by uncertainty over what government as owner wanted the corporation to be and do, by the old centralised structure of the BCNZ which prevented the operating divisions from getting on with the real job and the frustration of not knowing whether we were going to get a licence fee increase or not.

More radical reforms had to await the revolutionary deregulation of the industry and the splitting apart of the BCNZ brought by the government's new broadcasting policy. It must be remembered that from 1984 until well into the second half of 1988 the final structure of the television broadcasting industry in New Zealand remained unclear. It was only when the government started to deregulate telecommunications in 1987 that it became evident that radical change was also likely to take place in broadcasting. Up until this point all that was really known was that there would be a third channel competitor operating under the conditions laid down in the warrant issued by the Broadcasting Tribunal. It was this challenge that TVNZ had initially prepared itself to meet.

CHAPTER FOUR

Establishing TVNZ as a State-Owned Enterprise

The Labour government's policy on the structure of broadcasting was announced on 30 August 1988. Central to this policy was the decision to create two new broadcasting SOEs from the BCNZ, Television New Zealand Limited (TVNZ) and Radio New Zealand Limited (RNZ), as recommended by the Rennie Committee. This decision made clear the government's intention that TVNZ and RNZ operate in a fully commercial manner. The target date set for the commencement of TVNZ as a new SOE was 1 December 1988. This left only a short three-month period to decide how assets and staff of the BCNZ were to be divided between TVNZ and RNZ, valuations completed, financial structure determined, and the new SOEs legally established. The objective of this chapter is to discuss the events, processes, and decisions taken in this period to establish TVNZ as an SOE. However, before beginning this discussion we first provide a brief overview of the State-Owned Enterprises Act 1986 as necessary background.

The State-Owned Enterprises Act 1986

The philosophy behind changes made in New Zealand to government enterprises were clearly outlined in an *Economic Statement* announced to Parliament on 12 December 1985 by the Minister of Finance, Roger Douglas. In this statement the minister spelt out five key principles for the efficient management of public sector commercial activities. These principles then became the cornerstones of the State-Owned Enterprises Act 1986 and led to the deregulation of the markets in which the commercial activities of government enterprises took place.

• *Responsibility for non-commercial functions was to be kept separate from state-owned trading enterprises.*

Implementation of this principle meant that SOEs would be held accountable only for meeting commercial objectives and not for meeting social or non-commercial objectives. Policy and regulatory activities were to be transferred to other agencies, and mechanisms were provided whereby social objectives would be met by SOEs only through explicit contracting. The intention was to improve performance measurement and to avoid the confusion that existed in government enterprises as to the trade-offs between different objectives.

The contracting mechanism included in the SOE Act for this purpose was section 7, which states that

> Where the Crown wishes a State enterprise to provide goods and services to any persons, the Crown and the State enterprise shall enter into an agreement under which the State enterprise will provide goods and services in return for payment by the Crown of the whole or part of the price thereof.

Little direct use was made of this section. The government preferred instead to apply the contracting principle by establishing specialised purchasing agencies and by allowing direct arm's-length contracting between government departments and SOEs for the supply of goods and services. In the case of broadcasting, the government followed the recommendations of the Officials' Committee and decided to establish a Broadcasting Commission (now known as 'New Zealand on Air') as a specialised purchasing agent. Its function was to advance the government's social objectives in broadcasting by the purchase of particular types of programming—in particular adequate levels of universal coverage, local content, and minority-interest programming, with emphasis placed on reflecting and developing New Zealand's identity and culture. It was to be funded by a public broadcasting fee paid by the owners of television sets.

• *Managers of state-owned enterprises were to be given a principal objective of running them as successful business enterprises.*

36

This principle found expression in section 4(1) of the Act, which states:

> The principal objective of every State enterprise shall be to operate as a successful business and to this end, to be—
> (a) As profitable and efficient as comparable businesses not owned by the Crown; and
> (b) A good employer; and
> (c) An organization that exhibits a sense of social responsibility by having regard to the interests of the community in which it operates and by endeavouring to accommodate or encourage these when able to do so.

This principal objective provision was seen by those drafting the Act as placing a clear emphasis on what was required of SOEs. Its wording was also critically important to the separation of commercial and non-commercial objectives, the proper alignment of motivations and incentives, and the creation of an appropriate performance measurement and accountability structure that would largely overcome problems that had arisen in the past from mixed and unclear objectives.

The principal objective provision states unequivocally that SOEs must operate as successful businesses and, *to this end*, be good employers and exhibit a sense of social responsibility. Hence, the successful business requirement was considered to be the predominant objective under the SOE Act at the time TVNZ became an SOE. However, as alluded to in chapter 2, the issue of what was to be the predominant objective of TVNZ was to become a contentious matter, and we return to a discussion of this issue in chapter 7.

• *Managers of state-owned enterprises were to be given responsibility for decisions on the use of inputs and on pricing and marketing of their outputs.*

Under the Act SOE managers were freed to make commercial decisions without political interference. They were, however, to be held accountable for the results of their decisions. To this end the SOE Act set out accountability requirements in sections 14 and 15 of the Act. Section 14 requires the annual development of a statement of corporate intent (SCI) by each SOE and its review by shareholding ministers. The SCI is required to specify the objectives of the group, the nature and scope of activities, accounting policies, and performance targets. The SCI is important because it is, in essence, a contract between the corporation and its shareholders with regard to expected performance and a benchmark against which subsequent performance can be evaluated. To that end section 15 requires an annual report on operations; audited financial statements consisting of statements of financial position, profit and loss; and changes in financial position, and a statement of annual dividend payable.

Shareholding ministers are thus able to enter into an agreement with each SOE regarding performance targets and to receive information which enables them to monitor the SOE's achievement of those targets. In this way the government is able to monitor its investment in an SOE.

• *Individual state-owned enterprises were to be reconstituted on a case-by-case basis in a form appropriate to their commercial purpose under the guidance of private sector style boards.*

Under the Act each SOE would be a limited liability company with the Crown as its sole owner and with two shareholding ministers exercising these rights on behalf of the Crown. Government enterprises would be corporatised using the existing institutional governance structure provided by the Companies Act and the Commerce Act. This meant that each SOE would be governed by the general laws that apply to all companies.

In setting up each new SOE attention was to be given on case-by-case basis to the assets (and liabilities) to be transferred to the new SOE, the valuation of the net assets, and the structure of their opening balance sheets.

Like companies in the private sector, the governance of SOEs resides with boards of directors (but the directors are appointed by the government). Under the SOE Act, these boards are given very broad authority similar to that enjoyed by boards of directors of private sector companies. This includes the hiring and firing of managers and employees, the setting of management compensation and incentive schemes, oversight and review of major pricing and investment decisions, access to internal performance measures, and powers of internal audit.

Under section 5 of the Act the role of directors is clearly stated. It is to 'assist the State enterprise to achieve its principal objective' and 'to make decisions in accordance with its statement of corporate intent'. The board, not shareholding ministers, is responsible for operational decisions. The board is made responsible to shareholding ministers under clear accountability provisions which are set out in Part III of the SOE Act.

Under section 6 shareholding ministers are responsible to Parliament for 'the performance of functions given to them by the Act or the rules of the State enterprise'. In this respect boards answer to shareholding ministers for the consequences of their operational decisions in terms of meeting the SOEs objectives set down in its SCI. Shareholding ministers, in turn, are accountable to Parliament.

• *Unnecessary barriers to competition were to be removed so that commercial criteria would provide a fair assessment of managerial performance*

Many New Zealand government enterprises such as the BCNZ operated in an environment in which they enjoyed monopoly status or derived other benefits from state ownership. This created two problems: first, there was potential for misuse of this monopoly power with resulting losses to the economy; and second, without competition pressing on government enterprises it was difficult to assess managerial performance.

The principle of 'competitive neutrality' or 'the level playing field' was the result. SOEs when formed were to have neither special advantages nor disadvantages imposed by the state. Legislative monopolies and barriers to entry were to be removed and markets deregulated in order to achieve this result. In the case of broadcasting this would mean the removal of legislative and regulatory barriers to entry in order to allow new competitors to enter the market place.

With this background on some of the principles and provisions of the State-Owned Enterprises Act 1986 we turn now to a consideration of the actual processes used to establish TVNZ as an SOE in the short period of time from August to December 1988.

Trying Out a New Approach to SOE Establishment

Figure 4.1 provides a visual representation of the process that was used to establish TVNZ Ltd as an SOE. One of the first steps was to value the business to be transferred to the new company. It was important to get this value right to ensure that TVNZ Ltd would be faced with the need to establish adequate financial disciplines and that the shareholder would get an appropriate return on its investment.

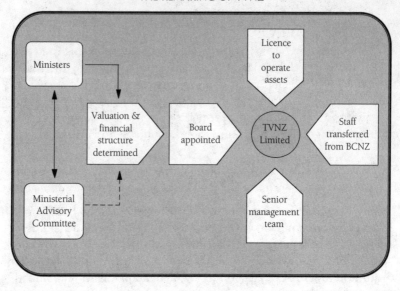

FIGURE 4.1: SOE ESTABLISHMENT

In setting up this process, Treasury officials were keen to avoid the protracted and often hostile negotiations over valuations and asset transfers that had occurred with SOE boards in prior waves of corporatisation. In the view of officials these negotiations had contributed to poor relations between shareholding ministers and boards, caused delays in restructuring, resulted in compromises from fair market values of assets, and led to the acceptance of financial structures for new SOEs that contained too little debt.

To avoid prolonged negotiations with the boards of the new broadcasting SOEs, the Treasury proposed a new procedure for valuing the assets and setting the financial structures of the broadcasting SOEs. As a Treasury official explained:

For TVNZ we looked at a model that had been used for Auckland Airport where there was going to be the possibility of a lot of conflict. The Crown was the major shareholder and there were a number of local authority shareholders. In this case a report was commissioned on what the value of the new airport company was and how it should be structured. This report was then used as the basis of an Order in Council which transferred all the assets by legislative means to set up the company. What they ended up with was not a sale and purchase agreement but rather a legislative instrument which detailed what assets were being sold, the price they were being sold for and the securities (equity and debt) that were going to be issued. That was simply a decision of the Minister of Transport in consultation with the Minister of Finance. There was no negotiation with the company. It worked out well and was put into place with very few problems and the thing seems to have worked well from then on.

The Treasury decided to use a similar procedure for the new broadcasting SOEs. The shareholding ministers would determine the asset valuation and financial structure. Prospective board members were not asked to negotiate the value at which assets would pass to the new SOEs. Rather their decision to accept an appointment to the board would be based on an asset valuation and financial structure that had already been determined against the background of the SOE Act. Although individuals invited to join the board were given full information on the valuation and financial position of TVNZ, this approach was quite different to those of the first two waves of corporatisation, where establishment (or transition) boards were appointed and played active roles in the valuation and the negotiation of an appropriate sale and purchase price.

This new approach was accepted by Cabinet, and the Minister of Broadcasting appointed a three-person Broadcasting Transition Board, comprising the chairmen-designate of TVNZ Ltd and RNZ Ltd and an 'independent' short-term director who would not be appointed to either board. The Treasury was represented in an ex officio capacity. The transition board was to be chaired by Brian Corban, the deputy chairman of the BCNZ and the chairman-designate of TVNZ Ltd.[1] The deputy chairman was Richard Rowley, another board member of the BCNZ and chairman-designate of RNZ Ltd. Rob Challinor, an executive director of the publicly listed bank, Bancorp Holdings Ltd, and a former partner of a national chartered accounting practice, was appointed as the short-term director.

The Treasury and the short-term director would have responsibility for advising ministers on the valuation and the financial structures of the two broadcasting SOEs. The chairmen-designate were to be consulted but would have not have any formal responsibility for either the valuation or the financial structure of the new companies. Treasury officials were hopeful that this new approach would avoid the incentive problems which had led to drawn-out negotiations with other SOE boards over the valuation of assets. The role of the transition board was to facilitate the physical establishment of the SOEs by advising ministers on matters relating to the division of assets and the allocation of resources, senior management appointments, and staffing.

Because the transition board lacked the authority to affect the final outcome of the valuation process, Brian Corban argued that the group should recommend to the Minister of Broadcasting that the transition board be called an advisory committee to better describe the status of the body. This recommendation was accepted by the minister and the first meeting of the Ministerial Advisory Committee (MAC), chaired by Brian Corban, was held on 4 October 1988. This left just eight weeks for the MAC to complete its work by the target establishment date of 1 December 1988.

The MAC seems to have functioned as an effective mechanism for quickly working through the many issues associated with establishing the broadcasting SOEs, including the appointment of management. In fact, given the compressed eight-week timetable (the most compressed of any SOE establishment with which we are familiar), it is highly probable that only a structure such as this would have worked in the circumstances. An internal Treasury memorandum written by an official involved in the process of setting up the broadcasting SOEs evaluated its operation as follows:

> The working of the MAC in providing an appropriate forum for the conflicts which had plagued other valuation exercises is [shown by] the fact that valuation and capital structure were agreed for both RNZ and TVNZ in a period of around 3–4 months rather than [those] for the protracted original SOE formations.
>
> Overall the exercise was conducted in a very positive spirit and was co-operative. There has been no major ill will created between the SOEs and the Treasury and in general the co-operation has continued in terms of the monitoring role which the Treasury has adopted.[2]

In the first half of October, the committee resolved to appoint Julian Mounter as chief executive-designate of the intended TVNZ Ltd subject to later confirmation by the new company's board. With this appointment, TVNZ got a chief executive who was already committed to the process of rebuilding it into an enterprise capable of surviving in a competitive environment. Because of the work that had already been done by the new chief executive-designate and his top management team, they were able to present business plans based on TVNZ's competitive strategy to the Ministerial Advisory Committee on 27 October 1988. These business plans were a primary source of data for the valuation exercise.

However, the process was not without problems. One important difficulty arose from a lack of clear decision approval authority on the part of the MAC during the brief period in which it operated. With a lame duck BCNZ board which continued to oversee operating matters and an advisory committee which lacked the executive authority of an establishment board, a decision vacuum resulted. TVNZ's top management had restructuring plans that needed board approval to go ahead but there was no establishment board to approve them. The new TVNZ Ltd board would not be appointed until the valuations were complete and accepted by ministers. In a letter to the committee dated 15 November 1988, Julian Mounter stated: 'TVNZ is faced with competitive pressures which demand rapid management reaction, within a timeframe that cannot be accommodated to the timetable of transition to an SOE.'

With TV3 due to go to air in June 1989 and a new pay television company on the way, Mounter was anxious to have TVNZ ready to meet the competition. He believed that time was of the essence and major decisions could not be delayed. Looking back on this period in a report to TVNZ Ltd's board in September 1991, Mounter wrote:

> TVNZ was at least two years behind the schedule I had set myself when I arrived and in my opinion we would lose the battle against the third channel, the economic battle to be truly profitable and the strategic battle to protect ourselves against the inevitable industry changes if there was not an immediate acceptance of major change.

In particular, Mounter asked for approval to move ahead on plans to restructure production activities so as to reduce expenditures. These plans included staff reductions, the external sourcing of drama production, the reduction of production in TVNZ's Christchurch station, and the shifting of some productions to TVNZ's Avalon studios to improve its utilisation for marketing reasons. In the end the best that the MAC could do was to give the chief executive-designate approval to proceed conditional on a confirmation from ministers.

Determining a Valuation and Financial Structure

The valuers, Fay Richwhite, chose to use a discounted cash flow

(DCF) analysis as the primary method of valuation. They defended this choice in their report as follows:

> The primary technique used by Fay, Richwhite to assess the value of the assets being transferred to TVNZ is to calculate the present value of the projected annual net cash flows of the operation in question using a discount rate appropriately adjusted for the risks inherent in television programme production, scheduling, networking, and transmission. This discounted cash flow (DCF) approach is particularly well suited to assets such as those involved in TVNZ's operations. It enables the particular circumstances facing TVNZ to be taken into account by modelling its response to deregulation, the major factors impinging on its investment, income and cashflow profiles over the next few years.

Forecasts of revenues and expenditures of TVNZ's businesses out to the year 2005 modelling TVNZ's response to its new and uncertain environment were developed. These projections were, in turn, based on forecasts using TVNZ's own extensive planning models. These forecasts were prepared by transition teams at TVNZ and in the Broadcasting Services Division of the BCNZ. For valuation purposes TVNZ's businesses were divided into three categories: (1) the New Zealand broadcasting business including the two broadcasting channels and signal distribution and transmission; (2) the non-broadcasting business of TVNZ including overseas sales of programmes, Avalon, pay-TV, overseas production and distribution, and a Pacific Rim satellite consortium; and (3) Broadcasting Services' new business ventures including the provision of service to TV3, lease of sites and antennae to other broadcasters, services to Sky Network channels, telecom activities, FM linking and co-siting, and general productions.

With respect to the transfer of contracts and assets from the BCNZ a 'warts-and-all' policy was applied. As with other SOE establishments, it was the view of Treasury officials that the process was more manageable if new SOEs were required to take on all assets and contracts of the predecessor organisation rather than being allowed to pick and choose. Where it appeared that the new SOE was being saddled with assets and liabilities it did not want, this could be allowed for in the valuation of those items.

While favouring a DCF approach, Fay Richwhite also prepared a valuation using conventional price–earnings (earnings capitalisation) techniques as a cross-check on the validity of the DCF valuation. However, the consultants questioned the usefulness of a price–earnings valuation in this case because of the volatility of TVNZ's expected earnings streams. In their report Fay Richwhite conclude:

> . . . one criticism of [a valuation based on P/E ratios] is that the profitability of TVNZ is expected to fall sharply after 1989 as competition from TV3 and other entrants builds up. . . . The only full response to this critique is to develop a full discounted cashflow analysis.

The valuation process is important not only to establish an appropriate value for the company but also to facilitate an improved understanding on the part of the leadership of the company's business strengths. In this respect, three important questions can be asked about the valuation process used in the TVNZ case. The first is whether there was sufficient challenge to and vigorous debate about, the forecasts and models of cash flows developed by management. The second is whether the process of valuation resulted in learning about the business by board members and managers. The third is whether the valuation provided an acceptable basis for measuring the future performance of the company.

Forecasts and Models of Cash Flows
The short timeframe for the valuation placed constraints on the level to which business plans could be developed and debated. However, given the time constraint there does appear to have been a reasonable level of challenge and review of management plans and assumptions by the short-term director and the valuation consultants. Major differences of opinion surfaced with respect to (1) the effect of competition on advertising revenues, (2) the timing and extent of capital expenditures needed to provide high-definition television (HDTV) services and maintain TVNZ's dominant competitive position, and (3) the need to switch to the distribution and broadcast of television signals by satellite. Other major uncertainties in the valuation involved the value to be placed on VHF and UHF frequency rights and the extent to which public broadcasting fee monies, now to be distributed by a broadcasting commission, would be available to TVNZ.

Because of high fixed costs, TVNZ's valuation was especially sensitive to variations in advertising revenue. An increase or decrease in forecast advertising revenue by 1 percent produced an 11 percent change in the estimated value of the business. Not surprisingly, there was considerable debate about the potential impact of competition on advertising revenues and, in particular, about the amounts that would be conceded to advertising agencies as commissions.

However, the most heated and important of the valuation debates between TVNZ and the valuers concerned the timing of the introduction of HDTV broadcasting and the associated capital expenditures. TVNZ's managers argued that unless it introduced HDTV in 1995 the new company would risk losing its dominant competitive position to competing HDTV broadcasts beamed directly into New Zealand by satellite. To retain its competitive position the company's provisional business plan provided for capital expenditures on HDTV of over $35m per annum over the period 1993 to 2007.

Based on their own independent review of likely penetration rates of HDTV sets and satellite dishes, the Fay Richwhite valuers argued strongly that it was unlikely that TVNZ would need to begin introducing HDTV broadcasting until the year 2000 or later. The valuers argued that the lack of any present definitive standard for HDTV and the low probability of one international standard emerging, suggested that full-scale introduction in New Zealand would be later than 1995.

The chief executive-designate and his senior managers vehemently disagreed with this assessment, producing evidence of a likely early HDTV introduction into Australia which may then be delivered directly into New Zealand through a direct broadcast satellite service. TVNZ's position was that a competitor's narrowcast system to even the limited number of subscribers who could afford new HDTV receivers would result in the loss of advertising revenue and could significantly affect the future profitability of the new SOE.

Because of the significant impact of the timing of capital expenditures for the introduction of HDTV on cash flows, several financial models were built: one which assumed increasing market share following introduction of HDTV in 1995; one for TVNZ's cash flow forecasts based on a deferral of the introduction of HDTV and a consequential loss of market share and revenue; and one which assumed HDTV expenditure deferred until 2000 with revenue forecasts reflecting satellite dish penetration reaching significant levels in 1998. Varying the assumptions underlying these three financial models yielded values for TVNZ of between $72.6m and $185.6m. This wide range of values reflected the uncertainty inherent in the new competitive and technological environment TVNZ was entering.

Based on their analysis, the recommendation of the Fay Richwhite valuers was that a gross value of the television assets be set at around $175m. The Treasury and the short-term director based their recommendations to the shareholding ministers primarily on the Fay Richwhite position on HDTV. Clearly, TVNZ's views on HDTV were not accepted. The values based on varying the assumptions underlying this financial model ranged between $170.3m and $185.6m. However, in a joint report dated 22 November 1988, Treasury officials and the short-term director pointed out that an earlier-than-expected introduction of the new HDTV technology would require an additional injection of equity into TVNZ by the government. Based on these considerations and final negotiations with management, a gross business valuation of $180m for the broadcasting assets was recommended.

This $180m value included all the broadcasting services (transmission and linking) assets which ultimately were to be vested in TVNZ Ltd's wholly-owned subsidiary. However, due to a paucity of information relating to prices to be paid for services provided by the new subsidiary, a separate valuation of these assets was deferred until the subsidiary commenced operations on 1 April 1989.

It is important to note that the scope of TVNZ businesses assumed in the valuation was wider than television broadcasting. As the valuation report to the shareholding ministers states:

> To achieve the targets upon which the valuations are based the businesses require flexibility to adopt the strategies forecast. These will include the ability to enter into joint venture arrangements and other risk diversifying strategies.

In this the valuers followed the recommendations of the Rennie Committee, who concluded that TVNZ would have to undertake a wider range of services in non-broadcast business, including programme sales overseas and international production and distribution, if it was to be commercially successful. The amount the valuers attributed to new business in the valuation was around $20m. A strong commitment on the part of management to a process of diversification into related businesses was assumed.

Learning about the Business by Board Members and Managers
The second question is whether the valuation process itself resulted in learning about the business by board members and managers.

Because board members, other than the chairman-designate, were appointed after the valuation was complete, learning about the business through active involvement in debates about future cash flows could not occur. This was an important loss and one thing that was traded off in this approach.[3] As Tipene O'Regan, a member of the last BCNZ board and the first TVNZ Ltd board, commented:

> I think a core of us should have been appointed, gone out and appointed a valuer and then agreed a value with the shareholder. I think the rest of the board would have had a better grip of the business if they'd gone through the modelling process.

Similarly, the corporate director of finance, Tony Gray, commented:

> If we did it again I'd advocate strongly that there should be a buyer/seller arrangement so that you get some market forces into the process. Part of the problem was that at the time the Ministerial Advisory Committee was operating nobody had been appointed to jobs in TVNZ. The chief executive wasn't appointed, I wasn't appointed—no one at that stage was appointed.

Learning did take place for those involved in the process as financial models were built and assumptions and business plans debated. In particular, the chairman-designate of TVNZ Ltd, the short-term director and the management of TVNZ learnt a lot about the future of the business from the valuation process. In mid 1989, after the valuation process was complete, the chairman Brian Corban recommended that Rob Challinor, the short-term director on the MAC, be appointed to the board of TVNZ Ltd as deputy chairman. He wished to ensure that Challinor's hard-won knowledge of the business gained through active participation in the valuation process was not lost to the board.

Performance Measurement

Lastly, the valuation must provide a reasonable basis for performance measurement. For this to happen it must be accepted by the board members and managers as a fair and acceptable basis on which to assess their performance. From our discussions with board members and managers it appears that the valuation and the associated financial structure were accepted as reasonable. Board members were given the opportunity to accept or decline appointment on the basis of a review of valuation documents. In this sense, at least, board members accepted knowing that the valuation would affect the measured financial performance of the company.

TVNZ's top management also accepted the valuation process as being fair, leaving the company with a strong balance sheet and a good start. Julian Mounter put it this way: 'When you are battling for something, every inch of ground seems important and there was so much at stake. But in retrospect it was a fair process and came out with fair judgements.' That this imposed valuation was accepted has a lot to do with the manner in which the short-term director approached his task and instructed the valuer. Rob Challinor explains:

> The first thing I said was this has to be a consultative process. We have to listen to management because they have already got their views and they're not going to be happy to manage unless they believe the valuation is correct. We need to bring management with us. So let's find out who has a view on the valuation and how the company should be run and listen to them. And when we reach conclusions, test those conclusions with them.

With the valuation completed, the remaining task was to recommend a financial structure for the new corporation and most importantly how TVNZ would be geared. Given the risk associated with TVNZ's future and the current need to upgrade and replace plant and equipment, Fay Richwhite recommended a comparatively low starting debt of $25m based on the recommended valuation of $175m. The Treasury and the short-term director concurred with this advice. They therefore recommended to the shareholding ministers that TVNZ Ltd assume existing BCNZ term debt of $25m on 1 December 1988.

Following extensive negotiations with management, provisions totalling $35m were recommended. These included $20m for redundancies in order to achieve cost savings in production areas; $6m for general restructuring costs including electronic news and computer systems and staff transfer expenses; $8m for being host broadcaster at the Commonwealth Games (to cover the shortfall between budgeted costs and anticipated revenues from foreign broadcasters plus intangible benefits derived from being host broadcaster); and $1m for outstanding legal claims. These provisions were subsequently reduced to $25m with the final structure approved by the shareholding ministers involving a gross valuation of $180m made up of debt of $15m, provisions of $25m and equity of $140m.

After the valuation was complete a last-minute problem arose.

Due to an oversight, working capital (stocks, debtors, cash balances less creditors) to be taken over from the BCNZ was overstated. On 5 December it was realised that working capital was $15–$20m less than the $39.11m included in the valuation because some residual liabilities had been overlooked (holiday pay owing but not taken, overdrafts in bank accounts, PAYE salary accruals and capital creditors). The problem was handled by an additional payment of $24.8m being made by the government to TVNZ Ltd after the establishment of the company. This problem with an under-assessment of liabilities reflected the relatively poor state of the BCNZ's financial reporting but also points to the need for careful assessment of estimates of liabilities in valuation processes.

Transferring Assets to the New Broadcasting SOEs

On the advice of Treasury, a two-step method was used by the government to effect the transfer of the BCNZ's assets to the new broadcasting SOEs. The first step involved transferring ownership of the assets to the Crown. This brought the assets within the ambit of the Treaty of Waitangi (State Enterprises) Act 1988 and section 9 (the Treaty of Waitangi provision) of the State-Owned Enterprises Act 1986. However, the intention was that the Crown's ownership would be momentary with an immediate transfer of the assets to the new broadcasting SOEs under an Order in Council as provided for by section 8 of the Broadcasting Amendment Act (No. 2) 1988.

Due to pressure of work at this time, the chief parliamentary counsel was unable to draft the required Orders in Council to allow the assets and liabilities to be transferred to the SOEs on 1 December 1988 as originally planned. Therefore, interim licences were given to TVNZ Ltd to operate the assets for a short three-week period to provide time for the Orders to be drafted and passed at the last Cabinet and Executive Council meetings in 1988. In the week before the last Cabinet meeting the Maori Council and the Nga Kaiwhakapumau i te Reo (the Wellington Maori Language Board) made an application to the High Court to prevent the transfer of the broadcasting assets, claiming that the transfer was in breach of section 9 (the Treaty of Waitangi provision) of the State-Owned Enterprises Act. The primary interest of Maori representatives in bringing this action was to protect the Maori language on television and radio.[4]

On the advice of their lawyers ministers agreed to stop the transfer pending discussions with Maori. In the interim the television assets remained in Crown ownership with TVNZ Ltd operating the assets under licence. Because the Crown retained direct ownership of the assets, the licence agreement placed constraints on TVNZ Ltd with respect to the sale and disposition of assets, its ability to borrow, and its ability to enter into joint venture arrangements. The original trial judge ruled in favour of the Crown allowing transfer of the television assets to TVNZ Ltd on 29 July 1991. This decision was appealed by Maori to the Court of Appeal which also ruled in favour of the original judgement on 30 April 1992.

The procedure designed to transfer the assets to TVNZ Ltd was not the method preferred by members of the Ministerial Advisory Committee or the management of TVNZ. To protect the broadcasting assets the chairman-designate argued for the direct transfer of assets from the BCNZ to the broadcasting SOEs. Because the BCNZ held title to its assets under the Broadcasting Act of 1976, which predated all Treaty of Waitangi legislation, the chairman argued that direct transfer of the BCNZ's assets and landholding would avoid the possibility of encumbrances being placed on the assets. Any claim would be against the old statutory corporation (which would have received cash for the assets that would ultimately revert to the Crown) and not against the physical assets of the new SOEs.

Treasury officials gave two reasons for advising ministers against this course of action. First, officials contend that if the transfer was done directly from the BCNZ to the new SOEs, ministers could have found themselves in a situation where the BCNZ board was negotiating asset values with the new boards of the SOEs with the Crown potentially excluded. Members of the MAC do not agree that this would have occurred. One member of the committee expressed his views this way:

> In my opinion it would have made no difference. It would have been done in exactly the same way. All we were doing was splitting one corporate into two. We said here is the BCNZ and this is the way the assets should be split up and this is the way it should be valued. Then corporatise it. I can't see how it would have made any difference at all. As it was the Maori claims caused huge problems for TVNZ because it was unable to sell off surplus assets as was assumed in the valuation.

Officials were also concerned that an attempt to directly transfer the BCNZ's assets would have been perceived by Maori as a deliberate attempt to circumvent the provisions of the Treaty of Waitangi (State Enterprises) Act. They therefore advised ministers that the Crown had an obligation to act in accordance with the spirit of the Treaty of Waitangi, that the likelihood of court action under a direct transfer between the BCNZ and the broadcasting SOEs was almost certain and that the Crown was likely to lose any resulting court action. Hugh Rennie, for one, still believes that Treasury's assessment on this point was in error and the claim that did emerge 'would not have arisen at all if the BCNZ's advice for direct asset transfers' had been followed. It would certainly have made life easier for the new broadcasting company.

A Mixed Board

Both Hugh Rennie and Brian Corban had concluded that TVNZ Ltd needed a small, commercially oriented board and made their views known to ministers. They both believed that it was important to avoid the large, representative boards that had been appointed to the BCNZ in the past.

However, in a break with past practice, the authority for making board appointments was delegated by the Minister for State-Owned Enterprises to the Minister of Broadcasting, Jonathan Hunt. In an apparent desire to ensure that issues involving the national interest, society, and culture were not forgotten in a drive for improved commercial performance, the minister pursued a mixed agenda in making board appointments. In addition to commercial acumen, issues of regional representation, gender, and ethnic mix were considered in making board appointments. One TVNZ Ltd board member commented as follows:

> My analysis is that [the minister] was seeking to achieve the best of the old, including his favourite programmes, but under the commercial imperative that the corporation should be self supporting. He was conservatory of broadcasting as well as accepting the need for change. He always had a fine sense of the importance of broadcasting to the New Zealand society and the difference between broadcasting's role in society and, say, the role of the Ministry of Works.[5]

As a result the board members appointed were chosen to provide regional and social balance as well as business skills.[6] To ensure social issues were not forgotten each board member received a letter of appointment from the Minister of Broadcasting which stated that within the context of being a successful business under the State-Owned Enterprises Act, TVNZ was to have 'the predominant objective of reflecting New Zealand's identity and culture and to encourage New Zealand programming'. This letter of appointment, given the requirement of the SOE Act, created a quandary for board members and led to contentious debate between the board and Treasury over the primary objectives of the new company. This debate clouded the relationship between the board and the shareholding ministers in the first two year of operations. We return to a further discussion of this issue in chapter 7. Here it is sufficient to note that the composition of the board and its method of appointment influenced an extended debate over dividend levels, profitability targets, and diversification and acquisition proposals. The nature of this debate was to raise serious questions in the minds of officials and shareholding ministers about the commercial orientation of the board.

To strengthen the commercial and financial skills available on the board, Brian Corban argued for the appointment of Rob Challinor, the short-term independent director on the MAC, as deputy chairman. This recommendation was accepted and Challinor joined the board in late 1989. Cliff Lyon, previously managing director of Wattie Industries Limited and a member of the State-Owned Enterprises Steering Committee, was appointed and attended his first board meeting on 13 August 1990. A further opportunity to change the composition of the board occurred when the terms of three board members expired at the end of 1990.[7] These members were not re-appointed and two new members were appointed to the board. They were Denford McDonald, managing director of Mitsubishi Motors Corporation and chairman of GCS Ltd, another SOE; and Rick Christie, a former managing director of Cable Price Downer. These new appointees had the reputation of being experienced and tough-minded executives who were familiar with companies operating in competitive commercial environments.

Asked for his views about the commercial orientation of the first board, Brian Corban responded:

> I would characterise it as a sufficiently strongly commercial board in the circumstances. The circumstances were that we had a very good chief executive. Julian Mounter was strongly commercial and he had a broad reach. There was sufficient commercial strength in Julian as chief executive, me as chairman and two or three other board members. We had seven faces around the board table some of whom were strongly com-

mercial and had wide experience and some who had very little commercial experience. But the end result was actually quite a strong synergy, providing that those who were not strongly commercial would not challenge necessary commercial decisions. I believe that a lack of commonality, a lack of homogeneity, a fair level of scepticism and constructive criticism is actually a powerful impetus.

But it also required a lot of effort on the part of the chairman in the initial period to hold such a disparate board together. We also asked two other members who remained on the board during this period to provide us with their assessment. The following comment is generally representative of the answers we received:

> I thought the initial board was pretty good. A lot of the issues that are important in television involve communication even down to the use of grammar so it was quite useful to have a couple of lay type people who were interested in that. Subsequently, the board was strengthened in a commercial sense by new members being appointed. There has been some suggestion that the first board was 'not commercial enough' but it was certainly a hell of lot more commercial than the board Radio New Zealand ended up with.

The Senior Management Team

As noted above major changes in the line-up of TVNZ's senior management team had taken place in the years leading up to corporatisation. Establishment as an SOE simply saw the top management of TVNZ transferred over to TVNZ Ltd.

First, Julian Mounter, who joined TVNZ in 1986 as director-general, was appointed by the Ministerial Advisory Committee as chief executive-designate; and his appointment was later confirmed by TVNZ Ltd's board. As director-general of TVNZ, Mounter had acted to revitalise and reorganise the senior management ranks of TVNZ bringing a number of new managers into the organisation. Making senior management appointments at the time of incorporation, therefore, was not a major issue as it was simply a case of continuing a process which was already well under way. Although a number of senior management positions were filled by managers who had been with TVNZ for a long period of time, a considerable number of new managers had been brought into key positions from outside. In particular, new managers were brought into the areas of finance and treasury, marketing and sales, programming and production, news and current affairs, and human resources management.

This process of bringing in new managers with commercial experience from outside the corporation continued with the hiring of additional managers such as Brent Harman (later to become general manager of TVNZ Networks and then chief executive on the departure of Julian Mounter), Des Brennan (who took over as director of sales and marketing), and Stewart McKenzie (who was brought in as director of finance for TVNZ Networks). The opening up of the organisation in this manner was an important factor in TVNZ Ltd achieving rapid organisational and cultural change.

Staffing the New Company

In the case of the BCNZ the government, rather than the new broadcasting SOEs, made the decision that all BCNZ staff would be appointed to one of the successor organisations. This mass transfer of staff, on the same terms and conditions, had the active support of the Public Service Association (PSA), the main union, as well as the more passive support of the the actors', musicians', and writers' unions as well as the internal Television Producers and Directors Association.

TVNZ Ltd had no option but to assume responsibility for staff already in TVNZ as a division of the BCNZ. This effectively precluded the organisation from using the change in its ownership structure and legal status to make a radical rationalisation of its staffing structure. As one senior manager recalls:

> We assumed responsibility for all staff knowing that the prospect was that a large number would ultimately be made redundant. The process of downsizing the organization has been the difficult part. If all positions had been vacated with all staff having to reapply for their jobs and all redundancies handled on day one, it would have been a different story. A lot of management changes had already taken place beforehand and the transition was a mass move which involved everyone in TVNZ and in parts of the BCNZ. The management team then had the job of restructuring the organization to make it a profitable business in a competitive environment.

As noted in the section on valuation a substantial provision was included in TVNZ Ltd's opening balance sheet to provide for redundancies and for general restructuring costs, as the government had decided that TVNZ should be responsible for any needed downsizing or restructuring of the organisation. An important problem facing TVNZ Ltd was changing the terms and conditions of employment of its staff. With competition on the way, it was vital that TVNZ should deploy its staff flexibly, rid the company of archaic work-practices, and be able to manage its pay structure.

Under the BCNZ, the industrial relations regime was characterised by a centralised bureaucracy with a 'career' service approach to staffing, linked pay scales between TVNZ and RNZ, a history of maintaining all of its operations and support services in-house, rigid job demarcation barriers, and strong union involvement with management. In this regime individual managers had little discretionary authority over wages or salaries, hiring and firing decisions, or who did what work when. Pay scales were specified in great detail along with numerous allowances and conditions, and job boundaries were carefully defined. Prior to becoming an SOE, TVNZ's industrial agreements set down rigid job boundaries covering 67 job classes and 214 job designations which specified exactly what each job did and did not entail. Multiple internal committees were involved in hiring, performance review, and promotion decisions and in the handling of disputes and complaints. Appeal procedures laid down in union agreements could delay new external appointments for months, raising the real possibility that good people would simply go elsewhere rather than wait for a decision to emerge from the arcane appeals process. In this environment, TVNZ's managers were unable to initiate any significant organisational change without extensive prior consultation with the unions. Doug Northey, an industrial relations manager at this time and later corporate director of human resources, recalled:

> In 1987 I went down on my knees to the union [the PSA] so that we could restructure the Avalon productions area. We had to have holy water put on by the union, otherwise they would use the machinery to lodge lots of appeals. So you needed their commitment to any changes otherwise they'd say 'We'll lodge 25 appeals if you do that, we'll tie you up in the public sector or local appeals machinery'.

Some progress had been made on industrial relations issues prior to TVNZ becoming an SOE. Most significantly, there had been a movement towards the use of individual contracts for on-camera performers, as was commonplace around the world. The use of individual contracts then spread outwards to include producers and journalists, who were making important contributions to the organisation but were being paid less than performers who had simply progressed upwards due to seniority. However, a more widespread movement towards the use of individual contracts had been constrained by a 1974 agreement between the BCNZ and the PSA to limit the number of staff on individual contracts to 7.5 percent of the workforce. Even though this agreement was not considered to be operative by the mid 1980s and was not legally binding, the real issue was that, in general, TVNZ's top management at this time would not stand up to the union or exercise the corporation's formal rights. TVNZ became more assertive about the use of individual contracts in 1987.

Progress was also made in removing some job demarcations and introducing multi-skilling. For example, a classic demarcation between studio and film camera, sound and editing operators was removed in 1987 following changes in technology which improved the quality of video tape making it comparable to the quality of film. New lightweight video equipment could be used outside where previously only film equipment could be used. Similarly, the artificial linkage between TVNZ and RNZ on staffing, wages and salaries, titles and positions was broken in April 1988 when TVNZ negotiated its first separate agreement with the PSA.

Establishment as an SOE, and the deregulation of broadcasting, allowed more flexible industrial relations and personnel policies to be put into place. First, TVNZ Ltd sought to increase flexibility by reducing the number of task designations written into job specifications, designations which not only restricted management discretion in the deployment of staff but also contributed to overmanning. Following corporatisation, the company refused to specify job designations in collective agreements and contracts, preferring instead to talk only about the 'size' of each job. Points were allocated to jobs according to the levels of skill, effort, and responsibility involved compared to those of other jobs. However, in the interests of retaining maximum staffing flexibility, even these job size definitions were not included in the collective agreements themselves.

In order to break the direct connection between seniority and pay rates, TVNZ Ltd introduced ranges of pay rates in place of specified steps within pay scales. Individual employees were placed within a pay range according to their experience and ability without the actual level being determined by seniority. The company also negotiated to eliminate penal rates and the multiple allowances for which employees were eligible. In a paper prepared in February 1992 the director of human resources, Doug Northey, observed that under the old regime:

> TVNZ had Penal Rates, Overtime, Shift Leave, Service Leave, Retiring Leave, and Allowances by the mile. Hours of work were excessively restrictive. It was even the practice to ensure that all overtime work was assigned to a week day, so that all weekend work was on Penal Rates, so maximising the cost to TVNZ.

Elsewhere in the broadcasting industry, flat rate contracts with few or no allowances were standard practice. While some progress was made in this area in the 1989–90 transition period by the new company, it was not until a new collective contract was signed on the eve of the passage of the Employment Contracts Act 1991, that a significant change in remuneration policy occurred. This new contract provided TVNZ Ltd with a much more flexible set of agreements. Employees were now expected to work between 4 and 10 hours a day for up to 10 days a fortnight without incurring overtime for the first 80 hours worked. New staff were also put on flat, hourly, or annual rates. Many existing staff also opted for flat-rate contracts in return for an increase in base-pay rates. By 1992 three-quarters of the company's staff were on these contracts. This provided managers with much more discretion over how they deployed their staff and improved their ability to control labour costs.[8]

Many personnel functions were devolved to business unit managers who were given ultimate authority over the hiring process and responsibility for salary determination and promotion decisions within guidelines laid down by the Human Resources department. Non-promotion and salary appeal processes that were available to existing employees were streamlined with an emphasis on 'front end' resolution of grievances and dissatisfactions at the workplace by line managers. At the time of writing there were still semi-formal salary grievance procedures in place but they were infrequently used.

Following the mass transfer of staff from the BCNZ, TVNZ Ltd has taken a measured and deliberate approach to changes in industrial relations and human resource management. As a consequence there has been very little industrial unrest since TVNZ Ltd was established as an SOE despite the significant downsizing and restructuring that has taken place since that time. The downsizing and restructuring of TVNZ Ltd is discussed in chapter 5 below.

Conclusion

The New Zealand State-Owned Enterprises Act 1986 set the framework for a new organisational form for the commercial activities of government enterprises. The SOEs established under this Act would be required to operate as successful and profitable businesses in new deregulated environments and within more comprehensive monitoring regimes.

Setting a valuation and financial structure, appointing the board and the senior management team, and staffing are the most important of a series of complementary and mutually reinforcing changes involved in the establishment of an SOE. Because of difficulties that had been experienced in prior waves of SOE establishments, a new approach was tried with the broadcasting SOEs. Generally speaking, this approach—which included, among other things, appointing a board only after the valuation and financial structure for the new SOE had been set—seems to have worked satisfactorily. While the downside of this approach was that most members of the new board missed the opportunity to learn about the business through the valuation and the process of negotiating a sale and purchase agreement, it may have been the only approach that would have seen the valuation and an associated transfer price set within the short three-month time period allotted to the task.

The new SOE was initially constrained by complications with the transfer of its assets which limited its ability to raise capital by selling assets. It was also compelled to accept the mass transfer of television staff from the old BCNZ, leaving it overmanned and facing the need for rapid downsizing. These complications created some difficulties for TVNZ Ltd in a very dynamic and potentially highly competitive industry.

CHAPTER FIVE

New Competitive Strategies

At the time TVNZ was established as an SOE the international broadcasting environment was going through a number of profound changes which were affecting the nature, shape, and economics of the industry. In particular, the convergence of transmission technologies (microwave, cable, fibre optics, and satellites combined with advances in digital technology and computer networks) meant that broadcasters and traditional telecommunications companies were facing both new competitors and new forms of competition. This convergence was reflected in the many mergers and alliances between major telecommunications firms, broadcasters, cable companies, satellite operators, owners of film and programme libraries, computer manufacturers, and software houses around the world.

Combined with the almost complete deregulation of broadcasting and telecommunications industries in New Zealand, these technological and industry changes meant that TVNZ was faced with threats and opportunities in both its television and signal distribution and transmission businesses. It was to these threats and opportunities that TVNZ responded in the manner depicted in figure 5.1. Once it had taken steps to reshape its core business and defend its market share in order to protect its revenue streams, it actively pursued a diversification strategy intended to broaden its base and ensure its long-term position.

Core Broadcasting Strategy

Defend Market Share

With the imminent entry of what appeared to be a formidable, well-funded broadcaster into their market, the first order of business was

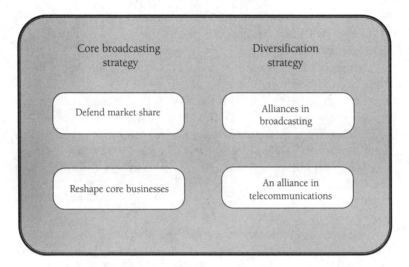

FIGURE 5.1: STRATEGIC RESPONSE

to develop and implement a strategy to defend TVNZ's market share and existing revenue base against its new competitor. The arrival of TV3 posed a real threat to TVNZ. It had strong backers and signalled its intention to take on TVNZ across the board. With its high fixed-cost structure, TVNZ's earnings and cash flows were extremely sensitive to even small changes in revenue. Any significant loss in revenues, therefore, could threaten the financial viability of the new SOE. Julian Mounter feared that if TV3 and other market entrants were able to win 30 to 40 percent of the market quickly, TVNZ would rapidly lose financial viability and face the prospect of being broken up. This was particularly a danger should the economy enter a downswing, as it did in 1990–91, where the total spent on advertising either remained stagnant or actually declined. A new competitor in a growing market is a different thing to a competitor entering a mature, no-growth market as television advertising promised to be in the early 1990s. In Mounter's view the situation required quick and drastic strategic action.

Win the audience. Led by Mounter, TVNZ's managers developed and implemented a strategy to win the audience through complementary programming of its two channels and a new approach to programme acquisition. TVNZ first sought to hang on to advertising revenues by reorganising and upgrading its sales and marketing

functions. Efforts were also made to co-ordinate its activities in programme scheduling, programme production and acquisition, and
sales and marketing.

To enhance the organisation's ability to win audiences and retain
related advertising revenues, Mounter persuaded TVNZ's senior managers that they had to exploit their two principal competitive advantages: their control of two nationwide television channels and their
ability to acquire and produce more top-rating programmes than
their competitors. The objective was to keep TVNZ's two channels
numbers one and two in the audience ratings and force TV3 into
third place.

To position each of the two channels, TVNZ's programmers
started with the placement of the news. Without a highly rated early
evening news programme, senior managers believed that Television
One would finish third in the ratings. To win the early evening and
to capture audiences for later prime-time viewing on Television One
and Channel 2, a decision was made to cancel the news on Channel
2 and concentrate all of TVNZ's news resources on Television One.
The news was also moved back half an hour to 6 p.m. Coming directly after the news at 6.30 p.m. would be a related current affairs
programme that would follow-up on some of the stories of the day.[1]
This would be complemented by a popular soap and a quiz or game
show on Channel 2, the objective being to split the viewing audience
with different tastes almost perfectly between the two channels, leaving TV3 nowhere to go but into third place.

Following the resignation of the director of programmes, Harold
Anderson, in mid 1989, the chief executive brought in highly regarded consultant programmers under contract from overseas to help
get the company's programme schedules and programming department whipped into shape before TV3 came to air. A director of the
advertising agency Saatchi and Saatchi commented that bringing
overseas programming consultants into the company 'just went
through TVNZ like shock waves'.

Research carried out by TVNZ showed that television audiences
divide into two primary categories: those who look for stimulation
and those who use television primarily for entertainment and relaxa-
tion. Television One was established as the channel for information
seekers attracting an audience that was more affluent, skewed towards males and the 40+ age group. Television One would carry
TVNZ's main news programmes, current affairs, drama, documenta-

ries, sports, and some good-quality entertainment programmes. Channel 2 programming was developed to attract entertainment seekers with movies, mini-series, soaps, game shows, quizzes, children's programming, and short news bulletins. Over time this process of complementary programming of the two channels has been further refined through extensive audience research and tracking of viewing tastes and preferences of different audience demographics. In essence, TVNZ Ltd moved to a two-channel branding strategy which was clearly evident in the on-air image and promotion of each channel as well as the type of programmes screened.[2]

Prior to the opening up of broadcasting to competition, programme acquisition was not an issue of strategic importance to TVNZ. As a monopoly buyer, TVNZ simply fixed the prices it would pay for each programme type and waited until prices fell to that level. This meant that New Zealand audiences often had to wait. As one manager told us: 'We didn't get *Roots* for a couple of years because we followed our pricing rules and said "Sorry, that is our price".'

The high cost of local programme production relative to the cost of acquiring rights to offshore programmes dictated that the bulk of the schedules of both TVNZ Ltd and TV3 must be made up of imported programmes. (Around this time rights to screen an imported overseas programme ranged from $5,000 to $10,000 per hour, whereas the average cost of all locally produced programmes was around $61,000 per hour. The cost of producing New Zealand drama ranged from $200,000 to $400,000 per hour.[3]) The extreme differences in these figures meant that, with competition, long-term access to popular overseas programmes and formats was going to be an important factor for competitive success.

The second leg of TVNZ's strategy was to ensure that the most popular overseas programmes and formats were available to TVNZ and not to TV3's programmers. As one senior manager put it: 'What you need to win in competitive and commercial television is programme rights. You don't need anything else, you just need programme rights.' TVNZ's programme acquisition staff systematically identified important international distributors and set about securing long-term renewable contracts and output deals. As a result of this effort, TVNZ Ltd was able to obtain the rights to many top-rating programmes from the United States, the UK, and Australia before TV3 came to air.[4] On 14 March 1989, the chief executive reported to the board that the company had acquired nineteen of the top twenty UK programmes, six of the top ten from the USA, six or seven of Australia's best, and had sixty-four mini-series ready for use. The strategy was to progressively strengthen the schedule by introducing new programmes in the run-up to TV3 going to air. In September 1989, the chief executive reported to the board that a review of the company's film acquisition lists showed that of the top ten all-time box office successes TVNZ had acquired nine with the tenth still to be assigned to television. The company also had ready a stock of nearly 700 first-run movies.

Sport was to be a main plank of TVNZ's strategy to win the audience. Exclusive rights to coverage of major sporting events would give TVNZ a competitive advantage in attracting viewers. Consequently, a lot of effort was put into acquiring rights to major national and international sporting events. As the national broadcaster for many years TVNZ had developed ties with important national sporting bodies including rugby football, rugby league, cricket, and women's netball. The company now sought to sign long-term, renewable contracts with the codes. Winning the right in mid 1989 to broadcast the 1992 Olympics was important to winning audiences and to lifting staff morale within the company.

The important downside to this change in programme acquisition strategy was its high cost. Average prices and stocks of programme holdings rose rapidly as TVNZ acquired popular programme series and formats and entered into output deals[5] and long-term contracts with distributors which gave the company first options on products. Intense competition between TVNZ Ltd and TV3 for top-rating programmes such as *Neighbours* and rights to broadcast major sporting events drove prices up sharply.

In September 1989 the chief executive advised the board that the cost of prime-time overseas product had risen by up to 500 percent for some individual programmes while the cost of top presenters had doubled. The costs of rights to sporting events skyrocketed. Programme stocks also rose significantly in 1989 and 1990, pushing up inventory carrying costs. Over the 1990 calendar year, in particular, holdings of programme stocks rose from $48.1m at the beginning of the year to close at $76.3m, almost a 60 percent increase.[6] Increases in prices and the costs of carrying programmes put great pressure on the company to make savings in other areas.

Even with such sharp increases, the cost of overseas programme rights was still considerably less expensive than the cost of producing programmes locally. In these circumstances why did TVNZ continue to produce local programmes? TVNZ's managers gave a number of reasons that go beyond simple considerations of short-term costs. First, they argued that screening locally produced programmes was an important way of differentiating TVNZ as an important local broadcaster. This was particularly the case with the basic core of TVNZ's local programming which includes news, current affairs, and sports programmes. Indeed, many locally produced programmes, including those in other genres such as drama and documentary (most of which were produced with funding assistance from the new Broadcasting Commission), were to rank among the top-rating programmes screened by TVNZ. In the longer run, it was seen as the principal way New Zealand free-to-air broadcasters could protect their ratings against international competitors which, given access to high-powered satellites, could develop the capability of beaming programmes directly into New Zealand homes bypassing terrestrial transmission.

The second reason TVNZ carried local programming was political. Allowing local programming to fall below some minimum threshold increased the political risk that the government would step in and introduce some form of local programme quotas. Third, there was a cultural reason. Making and screening local programmes was argued to be part of the culture of the organisation. Graeme Wilson, later to be general manager of TVNZ Networks, provided the following perspective:

> At the management level local programming has caused debate in the past but it is a fact of life working in television. If you consider that meeting commercial objectives is not simply maximising the return in any individual year but ensuring the viability, stability, and flexibility of the business over the long haul then you need to provide local programming. If you took the view that you would maximise profits in any one year by cutting out a whole range of programming involving New Zealand identity and culture you could make a ripper of a profit in that year but someone would either change the rules or you'd end up killing the golden goose in order to get one good golden egg. It is a balancing act and a difficult one. It is a very easy objective when you are a monopoly, an achievable objective if you are a dominant operator and bloody hard if it is competitive.

By playing on its strengths as the existing operator and by rapidly improving its programme scheduling, marketing, and promotion,

TVNZ won the initial rating battle with TV3. TVNZ's two channels quickly took number one and two in the ratings and, as planned, TV3 was forced into third place. As can be seen from the two panels of figure 5.2, over the period of our study TV3 remained in third place. However, the size of TV3's audiences increased as its transmission coverage improved. From an opening coverage of 59 percent of New Zealand households when it went to air in November 1989, TV3 continued to improve its coverage reaching 81 percent of households by 1991.

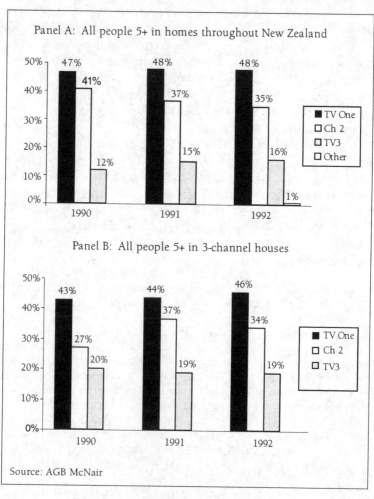

FIGURE 5.2: AVERAGE AUDIENCES 1990–1992

However, initially at least, TVNZ was assisted by the hang-overs of an excessive state bureaucracy and TV3's mistakes, as much as it was by its own strategy. After having been dragged through a long-drawn-out and exhausting application process and experiencing difficulty in raising finance, TV3 was late in going to air, making its opening broadcast on 28 November 1989, rather than on 1 June 1989 as originally planned. This delay provided TVNZ with the additional time it needed to get its own house in order and to let audiences adjust to its new programme schedules before TV3 got to air. Every week that passed saw TVNZ a little stronger and better prepared.

The initial delay in going to air also meant that TV3 went to air at the beginning of the summer season, when audience levels and advertising revenues enter the low point of their yearly cycle. To make matters even worse, TV3 went to air just before the XIVth Commonwealth Games in Auckland to which TVNZ held exclusive broadcasting rights. Not surprisingly, TV3's audience share in homes receiving all three channels, which stood at 19 percent on 24 January 1990 at the start of the Games, plummeted to a low of 7 percent by 3 February with TV3 averaging an audience share of just over 11 percent over the period of the games. TV3's low audience share, combined with its limited signal transmission coverage, had a considerable effect on its ability to gain advertising revenue and to generate needed cash flow. Had TV3 managed to get to air sooner it may have given TVNZ a better run for its money.

The two panels of figure 5.3 graph the number of programmes Television One, Channel 2 and TV3 had in the top 100 rating programmes in prime time (6.00 p.m. – 10.00 p.m.) over the first four weeks of competition in all homes and three-channel homes. TV3's initial inability to attract audiences away from TVNZ programmes to its programmes is shown quite starkly in these two graphs.

TV3's managers clearly underestimated TVNZ. They were stunned by the magnitude of TVNZ's response, which countered TV3's entry by having far stronger complementary schedules on its two channels,[7] extensive promotions of upcoming TVNZ programmes, advertising TV3's rating performance against the performance of their own two channels, and last-minute counter-programming against potentially popular TV3 programmes. A senior TV3 manager claims that TV3's problems were made worse because it was required, under the conditions of its operating warrant, to screen a high proportion of locally made programmes. This condition was not due to expire until

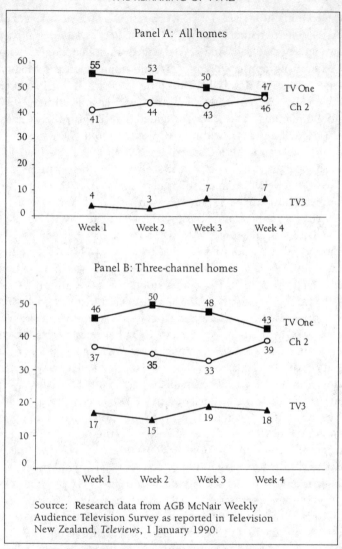

Panel A: All homes

Panel B: Three-channel homes

Source: Research data from AGB McNair Weekly
Audience Television Survey as reported in Television
New Zealand, *Televiews*, 1 January 1990.

FIGURE 5.3: PROGRAMMES RATING IN TOP 100
IN THE FIRST FOUR WEEKS OF COMPETITION

December 1992. No such condition was attached to the operation
of TVNZ.

TVNZ's managers believe that TV3 simply overreached itself, tried
to achieve unrealistic and unachievable targets, did not meet budgets,

and failed to develop a coherent strategy.[8] One TVNZ manager, who had worked for TV3 during its start-up phase, describes TV3 as being started by a group of very enthusiastic but naive people.

> You can see that from [TV3's] decision to be a fully fledged network with expensive news and current affairs and everything else from day one. I think this was a cardinal business mistake. They were spending money as if they had been in the business for five years and had a predictable cashflow. The psychology was that they were going to win and therefore they would be profitable in short order. They were hopelessly optimistic about beating TVNZ. At TV3 we had pep talks where the director of programmes said 'TVNZ have no idea about programming and I'm going to drive a wedge right through the middle of their two channels'. The general assumption was that TVNZ was moribund, a bunch of civil servants who didn't know about business, were not free-enterprise oriented and couldn't compete against the heat that TV3 was generating.

In reality, TV3 did not have a choice. A condition of its warrant was to have an independent news service. Nonetheless, Jeff Bennet, a senior manager at TV3 during this period, believes that even if they had a choice they would have started up with a news service because they wanted to compete with TVNZ across the board as a full-service network. In general, he accepted the preceding comments about TV3 during the start-up phase as being generally fair though somewhat simplistic. However, he did acknowledge that during this period 'there was a lot of hype, determination and optimism' in the company.

TV3 soon wilted under the weight of its difficulties, its mounting debt burden, the requirement on it to screen a high number of expensive locally produced programmes, and its inability to attract significant numbers of viewers away from TVNZ's programme schedule. As a result TV3's share price fell sharply and it went into receivership on 2 May 1990. Some commentators criticised TVNZ at this time for the strength of its response to TV3 and asserted that TVNZ used its strong market position unfairly to 'try to kill their competitor off'. In receivership, TV3 Network Limited filed a lawsuit against Television New Zealand Limited in the Auckland High Court, alleging TVNZ had breached the Commerce Act 1986 on a number of counts. At the time of writing this matter was still to be dealt with in court.

Speaking in Wellington on 30 September 1991 at an Institute of International Research Conference on broadcasting, TV3's first chief executive, Trevor Egerton, acknowledged that TV3's initial promoters and managers ignored some of the basic rules of new business start-ups. In particular, Egerton, pointed out that TV3

> . . . did not (or could not) (1) have modest audience and revenue expectations in its first year because of the short-term market return mentality of the 80's, (2) pitch costs accordingly, (3) have a capital base with plenty to cover a rainy day let alone a thunderstorm, (4) [had] very few principals [with] start-up experience, and (5) were forced to start in late November.

In December 1991, after 19 months in receivership, the TV3 network was reorganised and restructured with a 20 percent stake purchased by CanWest Global Communications Corporation, a Canadian television company. TV3's bankers Westpac Banking Corporation raised its shareholding to 48 percent with the remaining 32 percent held by TV3 Network Holdings. After TV3 was reorganised and restructured, it was able to improve its ratings performance from its disastrous start in 1990. While remaining in third place, TV3 began to achieve consistent audience shares of about 20 percent of the homes receiving all three channels. This represents a substantial segment of the market. Under its new owners, TV3 eventually achieved financial viability and thus remained a significant competitor of TVNZ for audience share and advertising revenue.

However, it is clear that TVNZ won the first rounds in its battle with TV3: it won the ratings war, retained a substantial share of the audience, and maintained its dominant market position. Its strategy of branding its channels succeeded and TVNZ had a good base from which to develop its entry into related telecommunications fields.

Transform sales and marketing activities. In a strictly commercial sense, television broadcasters are marketing organisations that market programmes to audiences and audiences to advertisers by means of a forward schedule. As part of the BCNZ, TVNZ was a monopoly provider of television advertising. Its salespeople simply had to wait for clients to walk in the door. It was an efficient order-taking organisation—the idea of providing service to clients did not appear to be high in its list of priorities. As a result TVNZ's sales and marketing personnel were not liked by advertisers, were resented by advertising agencies and were not well regarded within TVNZ. Their poor standing within TVNZ reflected the low appreciation of the commercial aspect of television. One senior manager described them this way: 'In the old TVNZ the sales and marketing department was a separate bunch of guys in white belts and shoes. You rarely saw them but if you did you crossed to the other side of the street to avoid them.'

Michael Dunlop, who was hired by Julian Mounter out of New Zealand's competitive radio market just prior to the establishment of TVNZ Ltd, was given the task of transforming sales and marketing into a client-focused and effective selling organisation. After talking with clients and studying television selling and marketing operations in other countries, Dunlop instituted a number of changes to the orientation and culture of the sales and marketing staff. These included bringing in new managers and appointing additional staff; forming small responsive teams who would be responsible for servicing particular groups of clients; instituting training programmes (to change the emphasis from order taking to client servicing and to develop needed selling, negotiating, and account management skills); and altering the manner in which advertising time was priced and sold.

The considerable investment made in restructuring and training sales and marketing staff paid off in improved relations with clients. Some advertising agencies who had long suffered TVNZ's arrogance and lack of service as a monopoly provider were impressed with the resulting changes. A senior director of Saatchi and Saatchi assessed the new company as having 'the most professional sales force in the industry which is streets ahead of the newspaper industry, better than radio and better than their competition'. To price and sell advertising time TVNZ had traditionally used a pre-empt pricing system. The price for a particular advertising time slot was set according to its estimates of the likely audience for the programme in its scheduled time. Based on how an advertising agency estimated the demand for slots on that programme, the agency could choose to buy in for their clients at 40, 60, 80 or 100 percent of the posted price. An agency buying in at less than 100 percent could be pre-empted and have to reschedule their bookings if another agency offered to pay a higher rate. Under this system TVNZ simply responded to demand. Dunlop sought a more pro-active approach which would allow sales and marketing to manage the demand for advertising slots.

The pre-empt rate card was dispensed with and a new fixed-rate card was developed.[9] TVNZ Ltd now set the price for advertising slots according to sales and marketing's estimate of the programme rating. Clients paid 100 percent of the currently posted price for an advertising spot but TVNZ had the right to raise or lower the price for the remaining spots in the programme. The price could be raised if spots sold quickly, or lowered if demand was slow. Clients who had already booked slots were protected against upwards movements in rates but also benefited from any lowering through a downwards adjustment in price. This allowed the schedule to be repriced regularly without disadvantaging those already committed to it. Dunlop comments:

> We constructed a new approach to television selling and put in a more sophisticated rate card [pricing] system which allowed us to control the market price based on demand. It was not done anywhere else that I know of in the way we do it. The key thing is that we anticipate demand for advertising spots on key programmes. Our objective was to capture a premium on the high rating programmes we had acquired.

In addition, advertising clients were offered increased discounts if they placed most of their advertising work with TVNZ's channels. As discounts increased in accordance with the amount of advertising placed there was a strong financial incentive for advertising clients to place a large proportion of their advertising with TVNZ.[10]

To encourage active selling and client servicing, sales and marketing staff were placed on an incentive-payment system. Instead of a straight salary they received a salary plus a bonus. Part of each individual's earned bonus (70 percent) was tied to the sales teams achieving their team targets. Another part (30 percent) was tied to the sales and marketing division as a whole achieving its group targets. Dunlop explained: 'I insisted on the group/team split because no single person in television sales controls the lot. You cannot pin it back to the individual so it is inappropriate to pay bonuses or commissions on that basis.' Des Brennan, a marketer from Anchor Foods, followed Michael Dunlop as director of marketing and sales. Bringing in a manager who had experience on the other side of the fence as an advertising client helped TVNZ to further improve on the client-service aspects of its sales and marketing efforts.

Link components of the value chain. Part of TVNZ Ltd's strategy was to co-ordinate and integrate its value chain of programming, programme acquisition and production, and sales and marketing activities more effectively.[11] Programme acquisition and production decisions had been typically made without systematic consideration of audience preferences. Programmers scheduled those programmes which had already been acquired or produced and sales and marketing staff acted as order-takers. In the language of operations manage-

72

ment, TVNZ's scheduling decisions in the past were pushed by production and acquisition decisions rather than pulled by audience demand.

An important change at TVNZ was the development of a demand-driven system for the co-ordinated management of programming, production and acquisition, and sales and marketing. This co-ordination did not happen overnight; it took time to develop. However, the objectives of the programmers were to deliver an overall platform of ratings to the sales and marketing department with emphasis placed on both the size and the type of audience. The key to achieving these objectives was audience research which became increasingly sophisticated after TVNZ became an SOE. Particular attention was paid to prime-time from 6.00 p.m. to 10.00 p.m. The director of programming, John McCready, explained how research was used to help set programme schedules:

> To give you an example, with Television One and TV3 both playing news the available audience is basically a younger, teenage audience so we try to place a programme on Channel 2 that is complementary to the News. This is *Neighbours*. We find the available audience using the AGB McNair measurements of how programmes rate and who is watching them. We also carry out research constantly with groups of people to find out not only what they watch, but why they watch it and what they think of it— are they positive or negative. All this information is computerised and helps us to plan schedules and look for programmes. We are always looking for the largest available audience. If it isn't big enough we may have to compete with TV3 and try to knock them off the audience they are going for with a better programme in that spot.

Based on an analysis of ratings information[12] and attitudinal research data,[13] programmers now planned generic programme schedules well in advance and developed and adjusted on-screen promotion strategies for upcoming programmes.[14] With a generic programme laid out, programme acquisition and production requirements can be determined. McCready explained:

> Programme acquisition used to be driven by the acquisition department which was quite separate from the programmers. Programmes were getting bought that programmers didn't even know about and film buyers were buying films without talking to the schedulers. Acquisition decisions are now driven by the schedule.

According to a former director of production, Don Reynolds, the same thinking was also applied to local programme production decisions: 'In production we now make programmes with programme slots in mind. We know where we are going to put them long before we start making them.' This was quite different to past practices which saw heads of production departments deciding what they would make, producing it and then 'tossing it over the wall' to programming on the expectation that if a programme was made then sooner or later it would be screened. Prior to corporatisation programme makers did not think in terms of a programme schedule. In the remade TVNZ, the balance of power shifted so that programming, scheduling, and marketing suddenly became more important than they had been. Indeed, programmers were now the key decision makers. It was no longer enough for producers to have a good idea; they needed to identify its potential audience, sell it to the programmers, and fit it to the schedule.

Audience research was also used by sales and marketing to price and sell advertising spots within particular programmes and to assist advertisers and agencies to plan cost-effective advertising campaigns. Media research, which provided information on advertising expenditures by media type, business categories, and individual clients, allowed sales and marketing teams to work more closely and effectively with advertising clients. This research also made it possible for sales and marketing to have an input into programming. John McCready again:

> To illustrate, sales and marketing do not want sport in prime-time because many of the advertisers who advertise in sports programmes will do so whether it is in prime-time or in fringe prime-time. Even though we may get a smaller audience at 10.30pm, sales and marketing will argue that we can earn more advertising revenue throughout the whole evening if we schedule it there. So we try to work with them without compromising the schedule. If they said we could sell a truck racing programme at 8.30pm and it didn't suit our programme line-up we'd say no.

Whatever happens, the economics of prime-time programming mean that total revenues are heavily dependent on the schedule. For this reason programmers must be extremely careful with that schedule. As the corporate director of finance, Tony Gray, explained: 'A one percent shift in the audience to the opposition during prime-time can cost TVNZ in the order of $3m dollars a year in revenues. That has to be kept in mind.' The reason for TVNZ wanting to retain first and second place in the ratings is obvious. The sensitivity of revenues to ratings also explains why TVNZ continued to spend heavily on determining what audiences want and on marketing and promoting their own programmes on screen.

Reshape Core Businesses

The first part of TVNZ strategy was to retain as much as possible of its audience share and revenue flows from its broadcasting activities. The second part was to adjust the size of its operations through downsizing and to reshape its core businesses in the interests of efficiency.

In its first year as an SOE, TVNZ Ltd's managers were generally preoccupied with preparing and positioning the company for the coming competitive battle with TV3 and preparing for the 1990 Commonwealth Games. Although the board approved reduction in staff numbers of 150 at their first meeting on 8 December 1988, management reportedly delayed immediate downsizing because they believed a substantial number of staff would be lost to TV3 when it went to air. TV3's delays in going to air caused this tactic to come partially unstuck. Consequently, a trade-off was made between achieving some cost savings and the potential impact of a loss of staff on TVNZ's competitive position. Consequently, staff reductions were not made as rapidly as was initially expected.

With the conclusion of the Commonwealth Games and some other important events and with the threat from TV3 contained, the company came under increasing pressure from shareholding ministers and their advisers to focus on ways to significantly reduce TVNZ's high fixed costs[15] by closing surplus facilities and by reducing staffing levels. In response, the chairman wrote to shareholding ministers on 26 March 1990:

> In view of the comments received by Ministers about the low level of profitability and dividend payout, the Board considered proposals for the downsizing and restructuring of the company. This is now being implemented and secures and improves our operating financial results for 1990 and subsequent years.

Large reductions in expenditures on news and sports followed. Regional news programmes were closed down, the formats of news programmes were changed,[16] and the amount of minority sports and daytime weekend sports televised were reduced. In addition, the director of resources, Rod Cornelius, was given the job of downsizing by cutting staff and closing down facilities around the country.[17] He explained that TVNZ's strategy was to alter its cost structure by turning fixed costs into variable costs:

> I worked closely with other directors and particularly the director of production to decide how many people we needed to cut and how we should move resources around. We decided to go to the lowest possible level in one cut. The idea was to reduce our operations to a base level and then hire in to meet peaks of production activity.

However, there was also concern at board level that management should make cuts in a sensitive way, that wherever possible people would have other opportunities and that needed creative and technical skills would remain available to the company to cover peaks of activity. The chairman, Brian Corban, explained his thinking and that of the chief executive in the following way:

> Both Julian and I could be pretty tough in the restructuring process, but equally we did not like the destructiveness that we saw in other restructurings. There are many ways of restructuring things. You can restructure like a butcher and look at what you've got left and say 'Was it worth it?' because you've actually destroyed more than you have created. I have never operated like that. I actually lectured the chief executive and the senior executives at the first meeting I had with them that we had to be very careful and very sensitive in how we treated people and the talent in the organisation.

In Corban's view, if they moved too quickly to downsize the company, they could destroy creative talents and technical skills that were important to have available to the emerging industry. With his background in his family's wine business Corban argued that 'when you deal with your own money, your own resources and your own employees, you do not consciously destroy things' and you 'don't do things that damage the industry' as a whole. In this context, Cornelius explains how he approached his task:

> We did a lot of different things: I set quite a lot of people up in jobs rather than just pushing them out the door. I set them up in companies, I leased them parts of our operations we didn't want anymore and I gave them part-time contracts to work for us. A lot are still in business out in the open market and are still competing for work from TVNZ.

Unless the restructuring was done with care, they believed some of the most creative and talented people in the company might simply move to Australia. It could then take the company (and the industry) years to rebuild productive capacity.

In assessing the speed with which TVNZ made cuts it is important to remember that the company, as a national television broadcaster, felt that it was under constant pressure from politicians, the press, and the public. Because of these pressures it made sense for TVNZ to take the time to assist other groups to take over and operate some of the more visible facilities that TVNZ wished to quit. Mounter offered this explanation:

> The best example of this was Christchurch. The regional stations were a heavy drain on profitability, but regional broadcasting is an important part of the [broadcasting] mix. Politically it is a highly sensitive issue. I was constantly concerned that we would be directed [not to withdraw], as we sought to withdraw from having regional stations of the size we then had (which often duplicated what we had at Auckland and Avalon). I would have had considerable sympathy with those who may have sought to prevent our withdrawal. For this reason, I argued that we should encourage a group to take over our studios and set up a local station, which eventually became Canterbury Television.

In his view, whether or not the new operators were ultimately successful was immaterial. They had the opportunity to be and the opportunity to provide real regional/local programming which, for a national broadcaster 'charged with everything from covering the Commonwealth Games to entertaining the whole of New Zealand nightly, was a diversion and a loss making one'.

Table 5.1 shows the effects of the company's downsizing.

TABLE 5.1: TVNZ STAFF NUMBERS 1988–1992

Month and year	Permanent staff	Full-time equivalent
June 1988	1,819	2,278
January 1989	1,742	2,205
January 1990	1,342	2,035
January 1991	1,129	1,779
January 1992	1,040	1,632
Source: TVNZ Human Resources Department		

The right-hand column shows full-time equivalent staff which includes fixed-term contractors, casuals, and overtime. A comparison of the numbers in these two columns shows how TVNZ moved towards a reduced and more flexible staffing structure with a higher proportion of staff on short-term contracts of one kind or another.

Related to its downsizing strategy were a number of important steps TVNZ took to redefine the boundaries of its core businesses. It did this by separating some upstream programme production activities as well as the downstream signal distribution and transmission business from the main broadcasting core. The upstream business units resulting from these boundary changes were South Pacific Pictures Ltd, Avalon Television Centre and a number of lesser business units such as First Scene and Moving Pictures. The main downstream business established was Broadcast Communications Ltd.

South Pacific Pictures Ltd. Unlike news and current affairs programming, which were produced every day using a standard format, the making of drama is complex, project-oriented work, subject to peaks and troughs of activity. As part of the BCNZ, TVNZ had built facilities and hired staff to cover peaks of activity, resulting in large overheads and long periods of time each year when resources devoted to drama production were under-utilised. Given the high fixed costs of producing drama in-house, a decision was made in 1987 to close the drama department and move its funding to a new, independent subsidiary to be called South Pacific Pictures (SPP).

SPP's primary mission was to provide TVNZ with New Zealand drama productions at substantially lower cost than could be achieved by either in-house production or by commissioning independent producers. To keep costs low, SPP was set up with only a few top-level staff on year-to-year contracts to manage the overall operation. Producers and production staff were to be hired on a project-by-project basis, an arrangement Julian Mounter had had experience with in the UK:

> I was on the board of Thames Television's subsidiary Cosgrove Hall which, like Euston Films, literally had 6 or 7 people at the top turning out large amounts of production by hiring people as they needed them and laying them off when they didn't. I was a great fan of this concept and so we introduced it here with South Pacific Pictures.

In essence, SPP was set up as a project management firm for drama production, sharing with the independent sector the ability to pull together project teams which are disbanded on the completion of a production. Short-term market-based contracts replaced longer in-house employment contracts, thereby buffering the core business of broadcasting from peaks and troughs of drama production.

To further lower costs and leverage their limited production budget, SPP was required to find at least 50 percent of its funding from outside sources such as co-producers or the Broadcasting Commission. To attract co-production money, SPP looked for drama and film projects which have both New Zealand and international appeal so that high production costs and risks could be shared off-shore. A lot of time was spent by SPP executives in searching out and developing relationships with potential partners and convincing them that SPP could deliver quality drama at a low cost. With a number of successful programmes under its belt, it became easier to attract co-producers to SPP projects. Having completed projects with a number of overseas production companies in Canada, Australia, and the UK, SPP targeted European Community countries, which were reputedly good payers for television drama products. SPP's future is helped by the fact that co-production deals are becoming more commonplace as television markets around the world fragment and production budgets are pinched.

Established as a separate subsidiary in September 1988, SPP became the country's largest producer of New Zealand-made television and cinematic drama. It is regarded within TVNZ as a successful venture, especially after its successes with *Marlin Bay* and *Shortland Street*. Between 1989 and 1992 it produced $107m worth of drama for television and cinema. Don Reynolds, who was a senior member of the SPP management team during the early part of this period and before that an independent producer, commented:

> SPP built itself up from nothing to a company which has a turnover of over $20m which is about the same as Avalon yet has nothing like the overheads. It has only a few permanent staff, very few assets and very little equipment. It is a very successful model which continues to provide TVNZ Networks with an ongoing stream of drama which is not fully funded by TVNZ. And it will give TVNZ added value in the future. Every time SPP makes another programme, the rights to that programme become an asset and over time they are going to be very valuable.

Avalon Television Centre. With the movement of the headquarters of TVNZ to the Auckland Television Centre, the closure of the drama department and a further decision to move to outside documentary production, the Avalon Television Centre was left with significant excess capacity. Built in the early 1970s with the capacity and capabilities to service the entire needs of television production and transmission into the next century, Avalon has 23,000 square metres of floor space. It is reported to be the largest television facility in Australasia. The problem confronting the new SOE was what to do with this expensive white elephant. Julian Mounter describes the dilemma they faced:

> If TVNZ had been in America, you'd have shut Avalon down because there would be many places to access the facilities they had. But operating in New Zealand, if we'd shut Avalon, we'd have had too few resources, because there weren't any other proper studios. So the answer was to isolate our redundant facilities at Avalon and try to fill them with work from elsewhere. I said to the board and to the staff at Avalon if it works they stay, if it doesn't it will close.

What this leaves unstated is the political row in Wellington that would have ensued if Avalon had been suddenly closed down by the company. Partly to avoid this possibility, a multi-faceted solution was tried instead. This kept available facilities that were needed by the company but which could not be fully utilised by it, enabled the Wellington independent production sector to continue growing, yet helped TVNZ to withdraw from Wellington and consolidate its operations in Auckland.

With a relatively moderate valuation placed on its assets at corporatisation, Avalon had a chance to succeed. It was moved outside TVNZ's core business and set up as an independent profit centre with a board of management made up of corporate managers or directors. Reg Russ, who had considerable involvement in the management of the change process within the old BCNZ, was appointed as general manager to restructure and reorient Avalon's business operations. The objectives were threefold: TVNZ wanted Avalon to be self-supporting; the company wanted a return on the $20m of assets at Avalon; and it wanted Avalon to be a source of cost-effective programming for its core broadcasting business.

To bring down fixed costs and gain a measure of flexibility, staff numbers were slashed and new personnel policies were put into place. People with particular skills such as film and tape editors and graphics people were put on fixed-term contracts, and other production staff were hired on a project by project basis, or at hourly rates in the case of receptionists or security guards. Russ explained his strategy: 'I tried to think of Avalon as an independent facilities or production house and I tried to model all our personnel and financial policies on those used by independent commercial operators.' Avalon was repositioned as a services and facilities house supplying recording, sound mixing, studio hire, props, etc., to the television, video, and film industry. Avalon's acquisition of the adjacent National Film Unit in March 1990 for $2.5m gave Avalon access to purpose-built film facilities, New Zealand's best film-processing laboratory, and its only licensed Dolby sound-mixing facility.

After a slower than expected start in attracting outside work, which was blamed on a general downturn in the economy in 1990, Avalon had a measure of success in bringing in business from outside the TVNZ group. By 1991 Avalon was bringing in one-third of its business from outside the company. Its primary income sources were television productions (mostly game shows and factual programmes); film production with international and domestic clients; contracts for the transmission of Television One (and Channel 2 in regional formats); commercial work involving corporate videos and advertisements; and the supply of location (non-studio) crews for television and film productions. In addition to renting facilities and services Avalon also invested in productions, sometimes by making otherwise under-utilised facilities and services available to co-production partners in return for a share of the co-production.

Avalon also started to move back into the film and programme production business, with the appointment of a director of production in 1992 to rebuild Avalon's programme production capabilities. The objective was to increase the utilisation of the facility and augment income earned through rental activities. In moving back into production Avalon had to tread carefully to avoid damaging its rental business by directly competing against its own clients for production funding. For this reason Avalon proceeded initially by forming alliances and joint ventures with writers and independent producers who had projects at various stages of development that could be produced using Avalon's facilities and services.

Smaller units. Other programme support services and resources were treated in a similar way. The Auckland-based internal design department, which has responsibilities for sets, scenery, props, and costumes, was set up as a business unit known as First Scene. The outside broadcast service units attached to the television operations in the four main centres were reorganised as a business unit known as Moving Pictures. The objective of these reorganisations was that the new business units should earn additional income from outside work to offset the overall cost of internal services. The corporate director of planning explained:

> Rather than absolutely downsizing everything TVNZ looked at the assets it had and said with facilities like Avalon, our design departments and our broadcast fleets we can probably make a good living if we open them to the industry as a whole rather than new operators coming in and setting up these facilities themselves.

In addition to these business units International Operations managed all satellite traffic requirements for the TVNZ group around the world. Utilising facilities in London, Los Angeles, Sydney, Melbourne and Auckland, it sought to defray part of its cost by selling downtime it had available to other satellite users. Satellite operations also managed a Pacific television service to a number of Pacific island nations.

Unfortunately, some of the new business units did not do as well as expected. In 1991 Moving Pictures, First Scene, and TVNZ International were all falling short of their financial performance targets.

Broadcast Communications Limited. As part of the government's broadcasting policy announced in August 1988, all downstream signal distribution and broadcast transmission assets taken over by TVNZ Ltd were to be vested in a separate fully owned subsidiary which was to operate at arm's length from its parent company. In response to this requirement Broadcast Communications Limited (BCL) was established as a subsidiary company from 1 July 1989 after valuation of its assets and the establishment of a financial structure.[18] The government's intention was that this network would not be monopolised by TVNZ Ltd but would allow the company's competitors access to the network under transparent contractual terms and pricing structures.

BCL's primary asset was its network of more than 500 linked high-altitude sites around New Zealand. Although the facilities on these sites were originally designed and developed to meet the linking and transmission needs of television broadcasting, they were also capable of meeting the needs of radio and telecommunications companies. First, BCL needed to replace its aging analogue microwave system with digital equipment at a cost of around $20m. BCL also sought to make its operations cost-competitive by reducing its staffing and by making other improvements.

To meet the requirements of government policy and to utilise more fully its capacity and capabilities, BCL began to sell its network services to other broadcasters. Although TVNZ remained its biggest client, other major clients included Sky Network Television Limited and TV3. In the case of TV3 it supplied transmission services and some linking services in competition with Telecom Corporation of New Zealand Limited. Canterbury Television, Radio New Zealand, the Totalisator Agency Board, the New Zealand Police, private radio operators, and a number of Pacific island states were among its other clients.

By supplying services to its primary competitor TV3, BCL and ultimately TVNZ Ltd gained from any increase in TV3's coverage. Indeed, the expansion of TV3's transmission area in 1990 to cover 80 percent of New Zealand was BCL's single biggest television project. Earnings on such work partially offset any losses of advertising revenues to TV3. Similarly, as Sky expanded its coverage, BCL gained as the supplier of services and TVNZ gained as a shareholder in both BCL and Sky. In each case TVNZ's position was partially protected by having two different relationships with each company. However, there was also a downside risk. When TV3 went into receivership it owed both BCL and Avalon considerable sums of money for services rendered. Thus, competitive success in its core broadcasting business created difficulty for those parts of its business outside its broadcasting core. Similarly, even though TVNZ stood to gain as a shareholder in Sky, BCL bore some of the start-up risk because it had to make additional asset-specific investments in transmission facilities to transmit Sky's three channels. Nonetheless, BCL has contributed significantly to the overall profits of the TVNZ group.

Ownership of the downstream signal distribution assets was strategically important because it enabled the TVNZ group to exploit opportunities presented by the rapid technological advances taking place in the deregulated broadcasting and telecommunications industries. We discuss these opportunities in the next section.

Diversification Strategy

Deregulation and the entry of TV3 meant that even if the absolute size of the market expanded, TVNZ would still suffer an inevitable loss of some audience share and advertising business. The company's 1990, 1991, and 1992 business plans all stress the point that, to remain financially viable, the company had to find ways to take advantage of the opportunities brought by deregulation of both broadcasting and telecommunications. Otherwise, it risked becoming a victim of fragmenting markets and competitors who sought to fill desirable gaps in these industries. With its free-to-air broadcasting business under pressure, the strategic issues for TVNZ were to find other sources of revenue, to protect its long-term access to programme rights, and to do what it could to forestall the entry of well-financed telecommunications companies into television broadcasting. One senior manager described TVNZ Ltd's business environment as follows:

> Traditional broadcasting is a shrinking business. Ultimately you can see a future in which this kind of activity won't take place at all because it will be replaced by other forms of electronic communications that are more attractive to people at home and which will allow them to deal with television in much more attractive ways. So the big job for TVNZ is to be in the right place in terms of the changes as they happen. Sufficiently ahead to get advantage but not so far out in front that you go down cul-de-sacs or get into expensive pioneering when you know getting into the slipstream of somebody might be the more attractive thing to do.

In accordance with this assessment TVNZ looked for expansion opportunities that would both complement its core businesses and allow it to learn about new technology. In most cases TVNZ sought to participate in these opportunities through strategic alliances and joint ventures in which they had an equity stake, rather than by seeking to expand the boundaries of the firm through horizontal or vertical integration. By leveraging the assets and capabilities it already possessed, TVNZ was able to avoid the need to raise large amounts of capital. By avoiding full vertical or horizontal integration, the company was able to conserve scarce managerial resources and avoid some of the management problems and risks associated with 100 percent ownership.

Some alliances and joint ventures have already been referred to. These include alliances and joint ventures between SPP and Avalon and domestic and overseas programme producers. Through these types of alliances and joint ventures, TVNZ sought to capture economies of scale in programme production.

Alliances with advertising clients were also engineered through sponsorship of major sporting events like the Whitbread around-the-world yacht race and the America's Cup. These sponsorships were designed so that all parties to the arrangement received a benefit: the sponsored sport was funded, advertisers received appropriate recognition and prime advertising time and TVNZ promoted additional spending on advertising through a tie-in arrangement with the client sponsor. These sponsorships were also aimed at creating an image for TVNZ Ltd and the sponsors, as companies which back and promote winners in New Zealand society.

But the most important alliances TVNZ entered involved those in broadcasting and in electronic communications.

Alliances in Broadcasting
Around the time TVNZ was restructured, the fastest-growing sector of the worldwide television industry was pay-TV. Unlike free-to-air television which relies primarily on advertising for its income, pay-TV relies on viewer subscriptions. For TVNZ an investment in pay-TV offered several opportunities. First, it offered a new source of income to the company. Second, because pay-TV would carry only minimal advertising or none, it did not directly threaten established sources of advertising revenues. Third, it would complement the company's free-to-air broadcasting activities and help TVNZ's managers learn about this new business. Fourth, the thinking was that it would help the company avoid being outflanked by new entrants who, in a deregulated market, were expected to fill any desirable niche that opened up in the New Zealand market.

The strategic question was not so much whether to get involved but how. Initially, consideration was given to TVNZ starting up its own pay-TV channel, but this was rejected in favour of spreading the high start-up risk by entering into an alliance with the strongest of its potential competitors. To this end TVNZ took a 35 percent stake in the new pay-TV company[19] called Sky Network Television Limited. Sky was officially launched on 8 May 1990 to offer a movie channel, a sports channel and a news channel.[20] Although it was known that Sky would take some time to become profitable, TVNZ's managers believed that it would ultimately be financially successful if properly managed. They had the advantage of being the first mover into New Zealand pay-TV and so the company decided to take a calculated risk.[21]

Starting out solely in the Auckland market, Sky progressively expanded its coverage to other areas of the country. Unlike Rupert Murdoch's Sky channel in the UK, which broadcasts direct to homes from high-powered satellites, New Zealand's Sky operated using encoded terrestrial UHF transmissions. Because Sky operated as a terrestrial broadcaster it contracted with BCL to supply it with transmission services. As Sky expanded its service to other regions throughout New Zealand, the transmission services supplied by BCL became an increasingly important component of the subsidiary company's income. TVNZ broadcasting operations were also assisted by joint purchasing and sharing arrangements for movies and some sporting events. Julian Mounter commented: 'The real benefit of being with Sky in the early days came from selling them our transmission capacity and through joint purchase deals. This reduced the risk of the investment.' Learning about subscription television was also expected to have longer-term benefits once Australia allowed pay-TV operators to operate in the Australian market. TVNZ attempted to position itself to benefit from any such opportunities that might present themselves in the Australian market. The company argued, unsuccessfully, that under the Closer Economic Relations initiative New Zealand should be exempt from an Australian proposal to limit foreign ownership of any pay-TV consortium to 35 percent.

TVNZ was forced to reduce its holding in Sky to 25.1 percent and then to 16.3 percent following investment of over $100m by four leading telecommunications and entertainment companies: Bell Atlantic, Ameritech, TeleCommunications Inc., and Time Warner.[22] This reduction in TVNZ's stake resulted from two events: first, a need for additional capital by the fledgling company and second, an amendment to New Zealand's Broadcasting Act which removed all restrictions on foreign ownership of broadcasting companies in New Zealand. Hearing about this change in government policy prior to the completion of negotiations for a holding in Sky, the new American shareholders made control of the company through a 51 percent holding a condition of their investment. To pick up this investor, TVNZ had no option but to sell down its share in Sky further than it had originally intended. However, it profited on the sale of its shares.[23]

Sky's losses were greater than expected in its first years of start-up operation and the company was disappointingly slow to reach break-even. TVNZ's financial exposure, however, was significantly reduced by its sale of over half of its stake in Sky in 1992.

Relationships between TVNZ Ltd and Sky were generally co-operative, as the original objective was to operate the two businesses in a co-operative and co-ordinated way in their respective areas of business. Indeed, an important benefit of the alliance with Sky was joint purchasing of programme rights in order to contain prices and to obtain broadcast rights at lower prices than TV3.

Early in the relationship there were difficulties in Sky's management and a senior executive, Michael Dunlop, was seconded from TVNZ to serve as managing director of Sky for a seven-month period. But there has also been at least one instance when Sky and TVNZ's interests have come into open conflict. Sky's attempt in 1992 to boost its subscriber base by tying up exclusive New Zealand rights to broadcast the All Black rugby tour of South Africa locked out the free-to-air broadcasters and their greater audience reach. This action was viewed by TVNZ senior managers as generally unhelpful to the relationship and resulted in a court action being filed against Sky by TVNZ. With the reduction in its holding to 16.3 percent, TVNZ lost one of its two seats on Sky's board, and consequently its ability to protect its interests through its shareholding was reduced.

In the early 1990s TVNZ also entered into a strategic alliance with one of Sky's other shareholders, TeleCommunications Inc., to launch Asia's first business television news channel. As managing shareholder TVNZ was responsible for the establishment and early operation of the business, which was to broadcast a channel known as Asia Business News by satellite in English (and later in other languages) to Hong Kong, the Philippines, Thailand, Indonesia, Sri Lanka, Taiwan, and Papua New Guinea. The intention was that distribution arrangements would enable it to broadcast to Malaysia and Singapore. TVNZ's ownership of a lease of a satellite transponder helped the company to form this alliance.

TVNZ also investigated, but ultimately did not proceed, with the purchase of interests in a number of other overseas programme production, distribution, and facility houses. The concern here was that the structure of the global television industry was changing rapidly. The power was shifting from programme buyers to programme suppliers as large international companies were taken over and merged to form vertically integrated multi-media conglomerates. Unless TVNZ diversified, Julian Mounter believed that it would find its access to the world's best programmes progressively reduced as broadcasting came to be dominated by giant media companies. Moreover, changing technologies and in particular the arrival of HDTV would require totally different production, transmission, and receiving equipment. Mounter argued that any broadcaster who had not raised its profitability to high levels and spread the load across a more diverse set of revenue-earning businesses risked being marginalised. In his opinion the BCNZ and many British and European broadcasters had been arrogant in believing that they could quietly 'stick to their knitting' because these changes were far off and would not threaten them. Although its relatively remote location and Australia's slowness to deregulate gave TVNZ a breathing space, Mounter argued forcefully to the board that the company should diversify into related areas to ensure its long-term survival.

A major opportunity presented itself in late 1989 when Bond Media Ltd, which owned Australia's Channel Nine network, proposed to sell a significant chunk of its equity to overseas investors. The company immediately set out to investigate the possibility of making a strategic investment, as Channel Nine was considered by TVNZ's managers to be the best television network in Australia. Julian Mounter believed there were a number of potential advantages. It would help to assure continued access to programme sources, allow for sharing of overseas facilities and staff, provide access to the large Australian market for New Zealand-made programmes, and help to protect the New Zealand market against future entry by this powerful Australian network. TVNZ's deputy chief executive, Darryl Dorrington, commented:

> The basic strategy was that we actually thought it made sense to try and develop a strong alliance with a top Australian broadcaster. In our view it could have achieved two things: it would have secured the relationship more strongly with Channel Nine and it would also have got us into a relationship with some of the other parties that would have made up the consortium to buy Channel Nine.

Although investigation of the acquisition of up to a 15 percent stake was tentatively approved by TVNZ's board, strong opposition from shareholding ministers and their Treasury advisers made it impossible for TVNZ to proceed with this proposed investment. The reasons for this opposition and the subsequent performance of Channel Nine are discussed further in chapter 7.

Some of the benefits which would have resulted from a horizontal alliance secured by an equity investment have since been achieved by agreements and contracts. TVNZ and Channel Nine share a news bureau in London and a satellite link across the Atlantic into Los Angeles, and TVNZ's Australian correspondent is based in Channel Nine's Sydney office. TVNZ also negotiated first right of refusal on all Channel Nine's broadcast programme material. Although a non-compete agreement between the two companies with respect to New Zealand was also negotiated, without the added protection of a strategic holding in Channel Nine, TVNZ had no substantial protection against Channel Nine (or any other Australian network for that matter) entering the New Zealand market. It was partly to gain some influence and a measure of protection against this possibility that a strategic investment in the Channel Nine network was considered in the first place.

An Alliance in Telecommunications
TVNZ Ltd's springboard into telecommunications was its ownership of Broadcast Communications Limited which managed the group's signal distribution and transmission assets. With the deregulation of telecommunications, TVNZ was presented with an opportunity to enter the telecommunications business through BCL. Worldwide convergence in technology had led TVNZ to consider itself more broadly as a telecommunications company rather than merely a broadcaster. In this respect Telecom New Zealand posed a major long-term competitive threat. TVNZ's emergent strategy was to confront Telecom on its own turf in the telecommunications field rather than wait for Telecom to enter broadcasting.

Originally BCL proposed that they build an alternative telephone company based on their existing network. This approach was rejected by TVNZ's senior management because they believed it would draw scarce management time and resources away from the broadcasting side of their business. Instead TVNZ decided to enter into a strategic alliance with Bell Canada International, a Canadian telephone operator; MCI Communications Corporation, an American long-distance telephone operator; and the Todd Corporation Limited.[24] The partners formed Clear Communications Limited (CLEAR). With over $100m of shareholder investment, CLEAR's sole task would be to set up an alternative telephone company to provide a full range of competitive toll, private line and specialist network services.

CLEAR was launched in November 1990. Each of the partners, including TVNZ Ltd, had a 25 percent shareholding. To limit its cash investment and its risk exposure, TVNZ negotiated the major part of its shareholding in return for the exclusive and permanent assignment of access to BCL's telecommunications system and the sale of BCL's existing voice, data and private line business to CLEAR. As with its investment in Sky, TVNZ stood to benefit from its involvement with CLEAR both as a shareholder and as a supplier of services. BCL had a substantial annual contract for the supply of telecommunication services to CLEAR, which uses BCL's digital microwave network and fibre optics capacity, together with its own switching, to provide local, long-distance and international voice and data services to its clients.

BCL and CLEAR agreed to plan jointly and co-ordinate the growth and development of BCL's network capacity to serve the needs of both television broadcasting and telecommunications. In their written agreement the two companies acknowledged that they have separate core businesses and agreed not to compete in those areas. Because of its importance to the TVNZ group as a whole, TVNZ's managers invested considerable time and effort to ensure a smooth working relationship between BCL and CLEAR.

Since its formation CLEAR generally performed close to budget with customers joining at a higher than expected rate. By the end of August 1992 CLEAR had gained over 12 percent of the national toll market. In the short term, problems with interconnect agreements with Telecom affected CLEAR's levels of profitability.[25] However, considerable new business opportunities are available in this industry, including cellular telephony and the provision of cellular transmission sites. In an attempt to gain a foothold in the Australian telecommunications industry as a springboard for entering more difficult markets in Asia, in 1992 BCL joined with Bell Canada, one of its partners in CLEAR and Channel Nine in Australia to form a further alliance to bid on work in Australia. The joint venture company, Horizon Telecommunications Limited, was used to bid for a substantial contract to build and operate a new telecommunications network for the New South Wales government. The other two bidders were British Telecom (to whom the tender was awarded) and Pacific Star, a joint venture between Telecom New Zealand and one of its share-holders, Bell Atlantic. With its alliance partners, BCL has continued to evaluate other opportunities in offshore markets.

The longer-term and strategic importance of TVNZ Ltd's holding in CLEAR was that it put the group onto a technological growth path which would give it the chance to capitalise on opportunities resulting from the rapid convergence of the telecommunication, computer and television broadcasting industries. In its role as an infrastructure supplier to both free-to-air broadcasting and pay-TV, as well as to the telecommunications industry, BCL was in a position to develop a unique understanding of the different needs of each of these sets of clients as convergence takes place. This would help position TVNZ in such a way as to enable it to remain central to developments in this ever-changing industry.

Conclusion

Formulating and implementing a new strategic direction to contend with fast-moving technology and the changing economics of television broadcasting in the face of the deregulation of television broadcasting and related telecommunications industries posed a particular challenge for TVNZ. Under the leadership of its chief executive and with the strong support of the board chairman, TVNZ's managers focused their strategic thinking first on developing a set of strategies in broadcasting in both marketing and production which would give them an unbeatable competitive advantage over its new free-to-air competitor, TV3.

A second plank of its emerging strategy was to look for growth opportunities that would complement TVNZ's existing capabilities and competencies in television broadcasting and electronic signal distribution and transmission. In this respect, the strategic direction followed by TVNZ was to seek economies of scope through diversification alliances in broadcasting and telecommunications.

TVNZ did not merely seek to defend its market share from the inevitable incursions of the new entrants, although such defence was a crucial part of its early strategy, as evidenced particularly in its branding of its two channels. It also actively sought out strategic expansion paths so as to position itself to benefit from future technological and market changes. In following both defensive and growth strategies, TVNZ exemplified both the defender and prospec-

THE REMAKING OF TVNZ

tor organisations of Miles and Snow's classic typology of strategic alternatives.[26] It sought to defend its existing market position at the same time as it sought out new expansion opportunities in related industries.

CHAPTER SIX

New Management Structures and Culture

To implement these strategies, TVNZ Ltd had to develop organisational structures, align its incentive and reward systems, and put in place processes to manage and control its core businesses as well as its strategic alliances. As illustrated in figure 6.1, the management challenge for the TVNZ group was to align its organisation structure, management systems and internal culture with its strategic direction in order to secure sustainable competitive advantage. In all of this a major task was to change the culture and attitudes of managers and staff and to infuse the organisation with new commercial and entrepreneurial values.

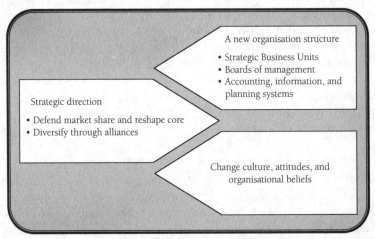

FIGURE 6.1: ALIGNING ORGANISATION STRUCTURE
AND CULTURE WITH STRATEGIC DIRECTION

A New Organisational Structure

Prior to corporatisation, TVNZ operated as a division of the BCNZ. It had a director-general who, along with the managers of other divisions, reported to a chief executive who was supported by corporate directors of finance, personnel and engineering and information technology. In August 1988, just prior to incorporation, the basic structure of TVNZ was as follows:

- *TVNZ management.* The director-general and a directorate made up of a director of resources, a director of programmes and production, a director of finance and a director of sales and marketing.
- *Resources.* A directorate responsible for personnel and engineering and support services in Auckland, Wellington, Christchurch, and Dunedin. Support services included regional administration; technical services involving electronic, film, sound, and vision equipment; transport; and property, maintenance, and security.
- *Programmes and production.* A directorate responsible for departments dealing with programme acquisitions (imports); and production including news, current affairs and sports; features; drama; documentaries; children's; entertainment; wildlife; and Maori programmes; outside broadcast events; and the operation of TV1 and TV2.

Described by one manager as 'a series of fiefs under the control of three titans', the major limitation of this structure was its failure to match responsibility with accountability for resource use. For example, although producers in the programmes and production department drove the demand for resources and facilities managed by the resources department, there was no system for holding producers accountable for the resources they used. As a result internal resources tended to be treated as a free good and used inefficiently. With the establishment of TVNZ Ltd as an SOE a new organisation structure began to evolve.

Forming Strategic Business Units

The first stage saw South Pacific Pictures Limited, Avalon Television Centre, and Broadcast Communications Limited separated from the core business of broadcasting. Each was set up as a stand-alone business with its own board of directors, or board of management if it was not a legally incorporated subsidiary. However, while this clarified lines of responsibility and accountability for the operation of these stand-alone units, they were not so clear within the core broadcasting business, which remained subject to the editorial and operational control of the chief executive and his corporate directors. This was not an ideal arrangement and created difficulties, as one manager describes:

> Initially, the core broadcasting business operated under the wing of the corporate directorate. Corporate directors were frustrated that it wasn't getting on and doing things and its managers were frustrated that corporate directors were interfering and getting in the way.

The second stage in TVNZ Ltd's organisational development began when Chris Gedye, a leased executive who had been appointed temporarily as general manager of human resources, put up a paper to the chief executive which proposed breaking the organisation into manageable business units to be known as strategic business units (SBUs). Gedye recommended a framework which would allow strategic and operational plans to be developed by discrete business units within TVNZ. Reliance would be placed on a market-based approach within the organisation through decentralisation, the use of profit centres, internal charging and results-based compensation.

The underlying principles of the proposed structure were straightforward and simple: (1) managers of SBUs would be given clearly identifiable objectives and would be held responsible for achieving specific results; (2) the total objectives of the organisation would be broken down among SBUs in such a way that all objectives had an 'owner' somewhere in the structure; (3) managers of SBUs would be provided with information necessary to control and monitor their performance against objectives; and (4) a primary role of corporate managers would be to monitor performance and to ensure that the actions of SBUs were consistent with overall corporate interests. To ensure that the manager of each SBU was held accountable for resources consumed in the pursuit of an SBU's objectives, a system of negotiated internal charges for resources, services, or products received from, or supplied to, other SBUs was to be put into place.

Gedye's proposed structure was accepted and a team of accountants was brought in under contract to design and implement the accounting and financial systems necessary to support the new structure and internal charging regime. Although the initial design of the structure and charging regime was progressively refined and modified over time, the SBU structure continued in place beyond the period under study. Figure 6.2 shows TVNZ's structure at the beginning of 1992.

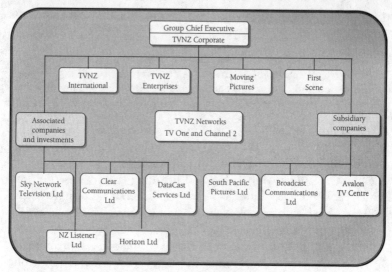

FIGURE 6.2: TVNZ LIMITED GROUP STRUCTURE 1992

- *TVNZ group.* The group structure was made up of TVNZ corporate, operating SBUs and subsidiary companies and investments and associated companies. TVNZ corporate includes the office of the group chief executive, the corporate directors of finance, human resources, planning and corporate affairs, resources and the company secretary/general counsel.
- *Operating SBUs.* Operating SBUs within the TVNZ group included TVNZ Networks, TVNZ International, TVNZ Enterprises, Moving Pictures and First Scene, South Pacific Pictures Limited, Broadcast Communications Ltd and Avalon Television Centre (which included the National Film Unit).
- *TVNZ Networks.* The largest of the SBUs within the TVNZ group was formed as a division of TVNZ Ltd in late 1990. The Network's division was, in turn, divided into a number of units which were also called SBUs for planning and budgeting purposes. These included: public relations, programmes, production and co-productions, natural history, news and current affairs, sports, sales and marketing, human resources, finance and operations and resources.

- *Associated companies and investments.* These included Sky Network Television Limited, Clear Communications Limited, DataCast Services, the NZ Listener, and Horizon.

In assessing the SBU system installed at TVNZ Ltd several points can be made. First, the term 'strategic business unit' as it was used within the company was really a misnomer as it was only the operating SBUs listed above which had separate and identifiable markets, strategies and boards of directors or boards of management. The one exception within TVNZ Networks was the natural history unit which was set up with its own board of management. However, putting nomenclature to one side, the general process of splitting TVNZ into separate business units provided a logical framework for planning and control that had been missing in the past. This was an important change. Through a system of internal charging, the SBU organisation gave each manager a 'bottom-line' which helped to focus his or her attention on alternative ways to achieve the objectives of their units. In this context, the SBU framework improved management processes and the structure of decision-making within TVNZ. The corporate director responsible for planning provided this viewpoint:

> The SBU system told people they were responsible and accountable for what they did in a way they were not when TVNZ was a public service monopoly. It made them pay attention to costs, to the quality of the service to their clients, made them more outwards looking and improved their management skills. It was a very valuable process for people to go through.

Whereas before producers would tend to use excessive quantities of a resource because it was not charged against their programme budgets, the SBU system of internal charging for all resources helped to change their behaviour. The first director of finance for TVNZ Networks, Stewart McKenzie, who was appointed in 1991, commented:

> The SBU system was important in getting understanding and accountability. In those terms it has been incredibly successful. There are still a few that say 'We're here to make good programming, we don't care about the cost of it'. But they are in the minority now.

However, actually getting the system up and running smoothly was not accomplished without considerable argument and conflict. Not surprisingly, the biggest fights were triggered by the internal charging regime. Initially, each SBU was set up with a balance sheet and charged by corporate for the cost of funds employed. Each SBU, in turn, was expected to fully recover all costs plus a mark-up to provide a return on its assets. The resulting internal charges brought an immediate and angry response from internal customers who argued that in many cases the new charges did not reflect market rates for equivalent services. From his time as a senior manager at SPP, Don Reynolds recalls:

> On our first major production at SPP I could get quotes from external suppliers of resources that were at least 70% less than what I would be charged to use TVNZ resources. All the internal prices were set on cost-plus basis and no one had gone out and looked at market rates. I started banging heads with people who said 'We know you can't do the job without us'. TVNZ people had so little experience outside TVNZ that they didn't understand that there was a whole free market industry out there that could provide anything.

The organisation's response was to permit managers to purchase services from outside providers in competition with internal suppliers. The intention was to drive down internal prices to reflect more closely market rates. Managers of internal resources replied that the cheaper outside 'market' rates were artificially low because external suppliers were temporarily shaving prices below full cost in order to survive the economic downturn in the economy. They began to lobby the chief executive to intervene and instruct SBUs to use internal services. However, he refused to intervene at this point and left it to SBU managers to sort it out among themselves. Don Reynolds again:

> I said to Julian 'How far do you want me to push this?' and he said 'Push it as far as you like'. So SPP produced its first two or three productions using few TVNZ resources, proving the point that drama could be produced without using TVNZ resources.

Market awareness certainly followed but so did some adverse consequences. The director of resources provided a classic example:

> Resources was required to make a certain target return on its assets so it simply put its prices up. I remember that one day Sports said 'The studio is too expensive for us to hire' so they did the programme out of their office on a Saturday. They hired staff in because they could get them really cheap from outside and they hired a camera. When we saw the programme on air it was appalling. We said 'Why did you do it?' They said 'Because it was cheaper'. I said that is utterly stupid. You just cost the

company double and you've produced a bloody awful product at the same time.

As a result of these problems a de facto company store policy emerged which, in the words of one manager, made it 'almost a sin to do work outside that could be done inside'. Julian Mounter commented:

> I asked Tony Gray [the director of finance] to run a 'second book' on some parts of the organisation while keeping pressure on the SBUs and every now and then I would intervene to tip the scales to logical conclusions. For instance: if the comparison with 'outside' leads to a reduction in staff and studios and a consequent saving, that is a proper result of normal commercial pressures, but if it resulted in you closing all studios and ending up having to hire outside for all studio work, you would have no influence on market prices and might have a subsequent increase in costs.

The internal charging regime was also reviewed and after heated debate the mark-ups on interest costs and requirements for a return on assets employed were removed. It was accepted that the internal resources SBU did not need to make a profit; it was sufficient that it broke even. This change in the internal charging regime reportedly brought the level of internal charges for facilities, resources, and services into rough alignment with market rates, and it virtually eliminated financial incentives for producers to hire facilities or resources from external suppliers.

However, this change in internal charging did not deal with a further complaint of programme producers. Producers argued that if unused capacity available in TVNZ's facilities and resources were charged out at marginal cost rather than full cost they could make more local programmes. This problem arose because of the way in which normal base charges for the use of internal resources were determined. Hourly charges were based on the total budgeted cost of a resource or facility (e.g., a studio or an editing suite) divided by the annual hours of programming planned. Some facilities and resources had capacity in excess of the projected volume of activity for the year. It is the pricing of this 'excess' capacity that was the bone of contention as there was no difference in rates for resource use within the planned volume of activity (i.e., within the business plan) and the incremental use of 'excess' capacity outside the plan. Consequently, producers had no financial incentive to make incremental programmes which were uneconomic under the current charging regime, but which would be economic and profitable to make if they were marginally costed. The internal charging dilemma was neatly summed up by one manager as follows:

> You could put all the resources cost into overheads and not include that cost in internal charges. It would then be just part of the overhead cost that needs to be recouped overall. The problem is that use of resources would then be free to all the programme producers who may block book facilities and then not use the resource. It would provide no incentive for people to use resources efficiently. The alternative is to say 'Everybody has to pay the going rate'. Then programmes don't get made because internal resources are too expensive.

Producers were told that where additional programmes beyond those included in the business plan were proposed and when some external funding was available, then that programme would be marginally costed and a decision to fund or not fund the additional cost would be made on this basis, but for accounting purposes the programme would continue to be fully costed. However, production managers were urged to make decisions that were in the best interests of TVNZ as a whole, even if this had a negative effect on their performance as measured formally by the accounting system. The finance director was reluctant to change the accounting system to allow for marginal costing of incremental programmes because he believed this would simply provide an opportunity for producers to opportunistically play the game by 'being the last cab off the rank because they know that is the one that will be the cheapest'. While this approach was organisationally untidy it did gain some measure of acceptance by programme and production managers. This is reflected in the following comment by the director of production:

> I've discovered that the more programming I push through our resources, the more programmes I am able to make. Many of the costs in television networks are fixed. I can drive down the costs per programme by pushing more work through these fixed resources or pushing independent producers who make programmes for us to use our resources.

The internal charging problems which existed within TVNZ Networks also existed between Networks and other SBUs such as SPP and Avalon.[1] Again the main approach taken to dealing with the problem was logic and persuasion. McKenzie commented:

Network producers will say this programme should take only $100,000 to make. If we put it at Avalon they'll charge us $140,000. We don't want to make the programme there. But from the TVNZ group perspective, because the facilities at Avalon exist and are not fully utilised, the incremental cost to the company may be only $100,000. My problem is to try to persuade people that irrespective of their internal charges it may be to the advantage of the group to make the programme at Avalon rather than making it outside the company.

Because well-made local programmes tend to rate highly and allow TVNZ to brand itself as a local broadcaster, TVNZ's managers wished to increase the amount of local programming available to its two networks. In some respects their SBU structure and the design of their internal charging regime got in the way of achieving this objective to the fullest extent possible. Moreover, there is a limit to how successful exhorting subunit managers to act in the broader interests of the firm (rather than their own narrow financial interests) can be. Alignment of evaluation and reward systems with strategic objectives is necessary to achieve desired behavioural changes.

There were other problems with the SBU system and internal charging regime installed by TVNZ which resulted from the transaction costs of running the system. This cost is a function of the extent of decentralisation and the volume of internal transactions. Although 27 SBUs were originally established, some aggregation and amalgamation later took place within TVNZ Networks, particularly in the Auckland Television Centre. Transaction costs can also rise dramatically if managers attempt to shift costs to other business units by charging for every service, no matter how trivial. A balance must be struck between using the SBU process as a means of changing managers' mind-sets and behaviour and the costs of operating the system. One senior manager made this point as follows: 'I think the SBU system was very effective as a change technique but it also created problems. You have got to be very careful it doesn't create its own bureaucracy by charging for everything.' The difficulties encountered by TVNZ are common to most internal charging regimes in the presence of under-utilised internal resources. They are problems which are inherent in the design of accounting and performance measurement systems which attempt to promote individual behaviours by business unit managers that are also in the overall interests of the firm. Managing internal resources and facilities for programme making is such a difficult, complex, and costly task that—we were told by a TVNZ manager—a number of Australian television companies have rid themselves of most of their production facilities. They now contract for the use of resources and facilities from outside suppliers as and when required.

It is interesting to note that in the UK the BBC has reorganised itself internally along somewhat similar but even more radical lines, by introducing something the BBC calls 'producer choice'.[2] This involves producers being charged for the use of internal resources. Producers are allowed to contract-out services if outside services provide better value for money. As was the case at TVNZ, the idea was to change producers' behaviour by making them responsible and accountable for their use of facilities and resources as well as to get a better indication of what programmes actually cost to make.

Boards of Management and Associated Investments

As legal subsidiaries BCL and SPP are overseen by boards of directors. This board structure was extended in the form of boards of management to other operating SBUs such as TVNZ Networks, Avalon, Moving Pictures, TVNZ Enterprises, and International Operations. Two or three corporate-level executives (corporate directors) and the senior management of the SBU sit on each board.

The principal role of each such board has been to monitor, oversee, and co-ordinate the strategies, plans, and activities of the subsidiary or SBU. The chief executive of the TVNZ group in 1992, Brent Harman, explained:

> In a company of our size and diversity, the reality is that the chief executive alone is not going to be able to monitor the achievement of SBU's business plans and help each manager with the problems he or she is going to encounter. I want each manager to look to the future and to grow the business. The role of the board of management is the same as a board of directors. I see them fulfilling exactly the same role as my parent company board does with me.

However, the system has not been without its critics, who claimed that it involved the expense of an additional layer of management at the corporate level. While conceding that the SBUs and related boards of management were not without problems, Harman defended their use arguing that they played an important role in the management of the group not only by helping to reduce spans of control, but also by providing a means of management training and cross-fertilisation within the TVNZ group of companies.

Considerable attention was also given to managing TVNZ Ltd's relationships with associated company investments, the most important of which is TVNZ's investment in Clear Communications Ltd. TVNZ's lead director on CLEAR's board was the deputy chief executive, Darryl Dorrington, who also chaired BCL's board. These related positions allowed the important interface between these two companies to be carefully managed. TVNZ's other director on CLEAR's board was TVNZ's group chief executive. A subsidiary, TVNZ Investments Ltd, held the shares in the associated companies Sky Network Television Limited, Clear Communications Limited, and the New Zealand Listener (1990) Limited. This separation helped to quarantine the risk associated with these investments from the core businesses of TVNZ Ltd.

Accounting, Information, and Planning Systems

Accounting and Information Systems

As noted earlier, the accounting and financial management systems of the BCNZ were considered by its chairman Hugh Rennie to be overly centralised, inadequate, and unreliable. This view was shared by the BCNZ's deputy chairman, Brian Corban, and the director-general of television, Julian Mounter. Even though these centralised facilities were viewed as too expensive and not appropriate in a competitive environment, TVNZ Ltd continued to use them until the end of December 1989. A contract to provide financial data was let with Computer Sciences of New Zealand using Works and Development Services Limited's computing facilities. This arrangement was found to be barely adequate as was reflected in the quality of the company's financial reporting over this period. Board documents reveal that some board members expressed their concern on a number of occasions over the poor quality of reporting and pressed management to move as urgently as possible to rectify the problem.

Having experienced the rigidity and comparative failure of centralised computing and the flexibility and success of their own diversified efforts prior to being established as an SOE, TVNZ Ltd's management opted to pursue a decentralised financial and information system strategy.[3] Provided they stayed within the overall framework of TVNZ group policy and standards, subsidiaries and major SBUs such as SPP, Avalon, and BCL were given responsibility for developing and operating accounting and other information systems which were appropriate for their particular activities. The same general approach was also followed within the core broadcasting business, where the finance directorate provided central accounting systems but business units were able to supplement this with their own information and systems. A policy of using off-the-shelf package software where it met at least 80 percent of TVNZ's requirements was followed. Computer development and support was handled by outside experts unless systems were considered to be strategically important (e.g., programming, sales and marketing and newsroom support).

The shift to a SBU structure resulted in the need for a reliable internal accounting system to track costs and expenditures (and sometime revenues) to both business units and programmes in order to measure performance against budgets. Each of the major SBUs (TVNZ Networks, Avalon, BCL and SPP) appointed a director of finance who, in turn, hired additional accounting staff. Within TVNZ Networks, for example, accountants were appointed to the most important business units to install specialised costing systems, to help business unit managers to understand the cost structures of their units and to ferret out slack in budgets. Stewart McKenzie, the director of finance of TVNZ Networks, indicated that this had helped to improve operating efficiencies:

> Since we put an accountant into news and current affairs and started tracking cost information, the efficiency of the news operation has improved and we are now able to make more news and current affairs programmes with the same amount of money as before. In the future we will start to compare our efficiencies in this area with international experience at say the BBC, or ABC and NBC in America.

To assist with the management of internal facilities and resources, the business unit responsible for facilities and resources installed a facilities management system (FACMAN). FACMAN was used to track projected and actual use of resources, to identify resources with available time and to report availabilities back to production units. As experience with using the system grew, resource managers started to explore questions about how pricing might be used to promote cost-effective behaviour on the part of programme producers. For example, news and current affairs block-booked facilities in order to ensure access in case of big story breaks. Resource managers began to consider a system of pre-emptive rights (with appropriate pricing) which would allow others to use the facilities but for news and current affairs to reclaim the facilities when needed urgently.

Important changes were also made to the manner in which programmes produced in-house were costed and how these costs and the cost of acquiring overseas programme rights, were accounted for. Under the costing system operated by the BCNZ, managers had no clear idea about what individual programmes cost to produce, as only direct (above the line) costs were charged to local programme production, and costs associated with the use of internal facilities and resources were written off as a cost of the period. This costing system had two effects: first, facilities and resources were often used inefficiently because producers and programmes were not charged for their use and second, programme inventories were under-costed, causing a distortion in reported profits. The accounting method used resulted in some programme-related costs such as salaries and expenses of staff working on the programme being expensed as a type of operational overhead in the period incurred rather than in the period in which programmes were screened and revenues earned.

To address these problems a system of 'total costing' was introduced. Programmes produced in-house were to be charged with the cost of internal facilities and resources used in their production. This was accomplished using the internal charging mechanism introduced with the SBU system. Programme-related costs were now charged to programmes and written off against revenue in the period the programme went to air, rather than written off to profit and loss during the period the expenses were incurred. The effect of this accounting change on reported profits is discussed in chapter 8.

The write-off policy on acquired programme rights (including movie rights) was also changed to promote good buying and scheduling habits on the part of programmers. This was an important change given the considerable increase in programme stocks in 1989 and 1990. As with the cost of in-house productions, the cost of acquiring programme rights was written off in the period the programme was played unless the right to more than one play was purchased.[4] McKenzie explained the policy:

> If we have rights to more than one play and we intend to play it more than once, then we will write it off two-thirds on the first play and one-third on the second play. We have also put teeth into the accounting

policy by having the computer kick in six months from the end of the licence period and start expensing any remaining programme cost at one sixth per month. We do the same with local programmes and in both cases it is charged against the budget of the programming unit. The director of programming is acutely aware of these costs flowing into his monthly report.

To assist programmers in the related task of programme acquisition and the management of their stock of rights, the company invested in the development of a programme purchasing and stock management system. The system operates on a schedule planned out several years and focuses on giving programmers the ability to manage their warehouse of thousands of hours of programme licences. By alerting programmers to second and third play options on all programmes, the system helped programmers schedule more efficiently. By identifying areas of their generic schedule where there were gaps and where they were overstocked it became possible for programmers to prepare shopping lists for their visits to television programme and film markets each year. This system was also being extended to allow schedule costing. The objective was to determine programming costs for different parts of the schedule, e.g., a night during prime time, a channel, a programme, and so on. When combined with information about ratings and advertising revenue earned by the schedule, it would be possible to determine gross contributions per night, per channel, or per programme. It should also be possible to calculate the cost per rating point. McKenzie explained how they expected to make use of this information:

> What we want to do is create an awareness about revenues and costs and where we make money and where we lose money. There is nothing wrong in losing money on a particular programme. We may choose to do that. But it is important that the programmers know what the costs of programmes in a particular schedule are. When they go to buy programmes they will have a better understanding of what they can afford to pay for a first run product for a particular schedule. With this information they will be able to say, 'I would like that programme for that slot but it is too expensive so I'll let it go'. It is easy to be seduced into buying programmes.

In this respect, the director of programming was now responsible for the management of all programme stocks, which were a key asset. Whereas each of the programme-producing units—news and current affairs, sports, natural history and children's productions—used to hold all their own work-in-progress and completed programmes, all these stocks were now centralised. This placed responsibility in the proper place—with the person who is ultimately responsible for all scheduling, programme acquisition, and utilisation decisions.

Sales and marketing activities were supported by specialised information systems which drew data on the availability of unsold advertising spots from the sales system and audience research data from the AGB McNair organisation. This information was then mixed and matched to assist sales staff to set selling prices and to develop selling tactics. A major development was to move this computing power into the field by equipping sales staff with laptop computers. This made it possible for sales staff to tap directly into the advertising spot scheduling system from the field to make critical sales. This was an important aspect of the company's strategy to meet competition in the sales and marketing area.

Planning Systems

Planning within TVNZ takes place at two levels: at a strategic level, and at an operational level where business plans are made. Strategic planning across the TVNZ group was primarily the responsibility of the chief executive and the corporate directors, assisted by a small strategic planning department. Business planning, on the other hand, placed more emphasis on both top-down and bottom-up processes.

For TVNZ Networks, the largest business unit, revenue projections were first made by the planning department using econometric models based on five-year forecasts of GDP growth, total advertising expenditure, and television's share of that total. These projections were then matched against a bottom-up projection, based on a client-by-client, sales-team-by-sales-team analysis. Differences between the top-down and bottom-up projections were then reconciled and a final forecast developed.

Attention then shifted to the cost side of the equation. The critical and variable factor here was the amount to commit to local programme production. As noted earlier, local programming was considered to be important to the longer-term profitability and success of the company. However, it was only after TVNZ's earning objective had been set for the year that a budget for the production of local drama, documentaries, features and children's and Maori programming could be determined.

Because local programming is expensive and is treated as a variable factor in the profits equation, TVNZ had some ability to manage its earnings in the short term simply by changing the amount it spent on local programming. However, as noted earlier, managers were acutely aware that too great a reduction for too long a time period would damage the company competitively and might provoke political intervention. Furthermore, Broadcasting Commission funding for local programmes was important at the margin to TVNZ's profitability and provided another incentive to engage in local programme production.

Once business units completed their detailed operational planning, these plans were then brought together by the planning department into consolidated business plans and individual plans reworked until an acceptable overall plan emerged and was approved. The management control cycle then started with the detailed budgets underlying business plans becoming the focus of the monitoring and control process. A review of the 1990, 1991, and 1992 plans showed a considerable improvement in the quality and sophistication in business plans produced over this period and, in particular, the development of more sophisticated five-year planning documentation.

The whole process of business planning, budgeting, and control introduced into TVNZ had an effect on producers and individual business unit managers. In the past producers had been allowed considerable latitude, and large cost overruns on productions were not uncommon. The introduction of better internal accounting systems and improved planning and budgeting processes and an internal charging regime brought more discipline to the process, allowing budgets to be challenged and slack to be rooted out. As one manager involved in planning put it:

> In the past producers would say 'Stuff the budget, I make beautiful pictures and I'll be damned if I am going to be accountable to a bunch of bean counters, I'm a programme maker!' The new finance director for TVNZ Networks just looked at them and said 'You are coming in on budget!' He knew he was making progress when the producers all started to carry little calculators to meetings.

Changing Culture, Attitudes, and Organisational Beliefs

The image of producers with calculators reflected the start of a fundamental change in attitudes and beliefs in an organisation that had for 25 years been controlled by programme producers on the crea-tive side and engineers on the technical side. An internal planning report written in 1987 referred to a number of characteristics of the organisation which would likely inhibit organisational change. These included:

> . . . the mediocrity that had been allowed to thrive in some production areas under 25 years of public service monopoly, the failure to hold people accountable for their performance, the practice of referring every decision upwards contributed to people's lack of feeling accountable and responsible, a lack of willingness to take risks, an in-built bias towards producing programmes in house, no encouragement of the development of new talent, a lack of research (information) to devise the best possi-ble programming strategy, a bureaucratic image and a cavalier attitude towards the public of 'take it or leave it'.

Injecting a more commercial, market-oriented focus into the organi-sation was a major challenge. In addition to the significant changes in TVNZ's environment brought about by deregulation and its estab-lishment as an SOE, several internal factors stand as influential in remaking TVNZ's culture and the attitudes and beliefs of its manag-ers and staff. Particularly important was the strong transformational leadership provided by Julian Mounter, with the support of the board, to implement needed changes.

As director-general of TVNZ and then as chief executive of TVNZ Ltd, Mounter was successful in persuading board members and managers that change in broadcasting was inevitable; that, in the fi-nal analysis, it was not being driven by legislation but by technologi-cal advances that were reshaping broadcasting and related telecom-munications on a global basis. His message was that the organisation needed to change the way it operated if it was to have a chance of survival.

Almost without exception the many board members and manag-ers we spoke with at TVNZ described Julian Mounter as a 'vision-ary' who clearly understood the direction broadcasting was taking on a global basis and saw a place for TVNZ in that future. The follow-ing are two representative statements taken from our many interviews with board members and managers:

> Mounter was a brilliant strategist, absolutely outstanding. He believed very strongly that TVNZ was a communications company and that free-to-air television would never survive in the long run. He believed that

TVNZ had to use its critical mass to tack onto [related] developments such as Sky and CLEAR and make money through them. I think he was the right man, at the right place, at the right time.

My personal assessment is that Julian's being here was the best thing that ever happened. He had no preconceived ideas and was able to look at everything afresh. He brought in a new broom and this was of immense benefit in turning the company around in a very short time.

The imminent arrival of new competition in broadcasting in the form of TV3 allowed Mounter to focus the minds of his managers, to quell opposition and resistance to the direction of change, and to attack the prevailing psychology of staff. In his opinion he had only a short time to convince people in the organisation that 'television everywhere in the world is war', and that if TV3 quickly won a third or more of the market, TVNZ would lose ground and might not survive. Mounter used the language of war to shock people into recognising that the market-place was not a sports field where to play the game is sufficient, but rather a battlefield where there are only winners and losers. In a videotaped speech to TVNZ's staff in August 1989 he had this to say:

The opposition is simply the enemy. There is only one rule and that is win, win, win. We are not in this business to come second but as we have two channels one of them will have to be second. We are not in the business to come third. That's their [TV3's] role. It's their name. They are the third channel. I am determined that this is where they will stay and I think they will find the opposition miserable, uncomfortable, uneconomic, untenable.

While Mounter's 'no holds barred—no quarter given' approach was successful in arousing competitive instincts in TVNZ, critics allege that on occasion it led to overly aggressive behaviour on the part of some TVNZ staff. Be that as it may, Mounter clearly saw the TVNZ's competitive situation as a useful spur to internal cultural change.

In 1987 Mounter led a series of planning meetings with senior staff. He recalls that time was of the essence:

I used to bully those meetings quite a lot more than one should have to. That is because we had so little time. It had to change quickly. In effect the changes were so big that had they been done slowly, they would have been defeated. The change had to take place quickly internally because the outside environment was changing so fast. If you don't do it quickly, you're going in a different direction by the time you finish.

It is not surprising that Mounter was viewed by some of his managers and staff as an authoritarian and combative manager who was hard on the company's managers and staff. As a consequence morale in the organisation was reported to be low for some time as changes were made. Managers who actively resisted change were moved aside and a number of new managers were brought into the organisation in key areas such as programming, news and current affairs, and sales and marketing (a number from the highly competitive radio industry). Other managers were reshuffled to match their abilities with the tasks they were being asked to perform. It was Mounter's view that in order to win against TV3 they had to change the attitudes of at least one-third of the staff prior to the arrival of competition and eventually at least one-half of the staff to ensure that the company wouldn't drift back into its old ways of doing things.

The starting point was sales and marketing. The objective here was to change the people who had direct contact with advertisers first. Considerable time and effort was expended accomplishing this before the arrival of competition. The thinking was that if the attitudes and culture of sales and marketing did not change rapidly, advertising revenue would be lost and the company itself might be in jeopardy. The other key change was the shift of power to programme scheduling and marketing and away from programme producers. As the environment changed, producers inside TVNZ and independent producers on the outside had to change rapidly in order to survive. Roger Horrocks, a member of the new Broadcasting Commission, commented:

> Basically to stay in business programme makers have got to know what broadcasters want. Just as the Commission knows what broadcasters want. Secondly, [programme makers] must understand programming and thirdly, [they] have to understand how to make a programme people want to see.

Some programme makers failed to make this shift in thinking and remain bitter about the changes in orientation and culture that occurred as TVNZ was transformed into a successful business enterprise.

While the impending arrival of a new free-to-air competitor was used by Mounter to concentrate the minds of managers and staff, the establishment of TVNZ as an SOE provided the organisation with the flexibility needed to formulate and implement strategy and plans with a minimum of outside interference. It also provided access to many other tools and mechanisms that could be used to promote cultural change. Perhaps the most important organisational innovation in TVNZ Ltd was the introduction of SBUs and associated internal charging mechanisms. Together with improvements in accounting and budgeting systems, the SBU system was instrumental in moving changes in culture and attitudes deeper into the organisation. To change culture, you need to change the expectations and motivations of managers and staff. The new SBU structure, with its associated internal charging regime and concern for the bottom line, required even producers to modify their behaviour. Over time, changes were expected to impact on the attitudes and beliefs which underpinned the structure.

It is important to understand that SBUs were not Julian Mounter's idea. Rather it was Mounter's hard-driving management style and his desire for change which, combined with competition and SOE status, created a climate in which initiatives could be taken by other managers. Mounter's contribution was to create an environment in which managers could manage. As the first director of sales and marketing appointed by Mounter recalled: 'Julian put in place the processes and said do it.'

Even his competitors admired what Mounter was able to achieve at TVNZ. A senior executive involved in the start-up of TV3 commented:

TVNZ changed their whole corporate culture in the time we were delayed in getting to air. They did an excellent job. While I don't particularly like Julian Mounter I admire the job he did. I also didn't like the way he did it but the result was exactly what TVNZ needed. Based on the way they have turned out you would have to say that it is probably a model SOE.

Hugh Rennie agrees. In correspondence with the authors he wrote:

My own assessment was that as an SOE, TVNZ had only about a 60% chance of business survival and success. TVNZ were relentlessly effective in achieving executive change, policy review and the building of a true commercial organization. In contrast Radio New Zealand (which I considered had a 90% chance of survival and success) failed to make the changes needed.

Finally, some attention was given within TVNZ to formal programmes aimed at promoting cultural change. The first of these was a series of seminars called *What If I Owned the Business*. This programme introduced groups of staff to the concept of competition, made people think about where the business was going, and with the SBU process helped make employees aware of things that were important to their parts of the organisation. Thus, instead of resisting change, people were motivated to redirect their energy into implementing change. The message was that if they didn't win out against TV3, it might be fatal for the company and for their jobs as well. In a report to the board dated 12 June 1989 the chief executive noted that about 1,000 staff had attended these seminars in the four main centres.

This emphasis on change and growth continued with team briefings (judged by some managers we interviewed not to have been particularly successful), and later a team was established to oversee a project called 'Vision 20/20'. The task of this team was to ensure all employees of the TVNZ group were informed of the changes taking place and envisaged in the future of the television industry. The team held a series of seminars attended by staff in groups (each group comprising employees from different units across the group) at which the challenges facing the company were openly discussed. Videos were shown, including some especially prepared for the sessions. The objective of these videos was to communicate a vision for the future of the industry and to provide a focal point for the seminars. The concept behind Vision 20/20 was also extended to researching specific topics, establishing a data base and providing information to managers and staff about future directions in the industry.

Without extensive longitudinal surveys it was difficult to gauge how far changes in culture and attitudes had progressed in the company. Julian Mounter provided this assessment in 1992:

> I thought it would take 10 years to really change the company as a whole and I still think that is right. I would say that currently 51 to 52% of the staff are real subscribers to the change and I think there are a lot of non-subscribers who have decided to go along as fellow travellers and not buck the system. However, if there was a hint that they would have support for going back to the old days of protectionism and everything else, they would drift back.

One staff manager, whose job brought him into contact with many parts of the organisation, provided the following perspective on the extent of cultural change:

At the swimming pool I use there are warm patches of water and cold patches where there are shadows. TVNZ reminds me of that. There are cold patches where the old civil service is strong and there are warm patches where the people are much more entrepreneurial and have taken on a new set of values. News and current affairs people were very market oriented but they have always been like that. At the corporate level there was intense rivalry with TV3 and then down in the middle, particularly in production, there is still this big block of civil service people.

In late 1991, Julian Mounter resigned from TVNZ Ltd. Some managers and outside commentators believe that by this time he had started to lose his effectiveness in the company and decided it was time to step down. This was not the first time Mounter had considered leaving the company. Indeed, he had expressed a wish to step down some 18 months earlier but had been asked by the board chairman (with the support of the board) to stay on.

Brent Harman, then general manager of TVNZ Networks, was appointed as group chief executive. A New Zealander, Harman was initially brought into TVNZ Ltd from the radio industry where he had been general manager of the radio station 1ZB. While Harman agreed that the strategic direction established by Mounter was correct, he pointed out that he had a different management style and placed a greater focus on management systems and on human resource management than did his predecessor. Harman explained:

> I've got a different style. I prefer to create the right environment to let people get on and do things. Being able to do that reflects the stage of development the company is at. Julian had hands-on control because when the company was going through a lot of change and adopting a new direction, the [chief executive] had to sit there as puppeteer, otherwise it could take off in the wrong direction. Now the direction is fairly well set, so the key managers can [be left] to get on with it.

Mounter was necessary to lead the radical organisational change required by TVNZ's establishment as an SOE. He was a forceful, transformational leader at the top of the organisation. Once a new strategic direction was established and the organisation had been turned around and put on a commercial track, what was required for the longer haul was a more transactional style of leadership that paid attention to getting the right systems in place and empowering the management team. Harman promised that style of leadership to TVNZ in contrast to Mounter's more charismatic, top-down approach.

Conclusion

A combination of factors worked together to influence organisational and cultural changes within TVNZ. These factors included:

- deregulation of broadcasting and the imminent entry of competition into the television broadcasting market,
- establishment as an SOE,
- the leadership and strategic insight of the first chief executive of TVNZ Ltd supported by the board and its chairman,
- changes made in the ranks of senior management,
- introduction of an organisation structure designed around SBUs and related improvements in accounting, planning, and control systems,
- progressive introduction of an industrial relations regime which allowed more flexible personnel policies and work practices,
- introduction of new 'pay-for-performance' incentive mechanisms in sales and marketing and
- implementation of explicit programmes of cultural change.

While it is the combination of all these factors that drove change in TVNZ, two stand out in terms of importance. The first was the establishment of TVNZ as an SOE in the context of the deregulation of the broadcasting and related telecommunications industries. The second was the critical role played by the chief executive, Julian Mounter, with the close support of the chairman of TVNZ's board, Brian Corban. Together the right environment and the right people paved the way for dramatic and radical organisational change.

CHAPTER SEVEN

Relationships and Pressures

We turn now to a review of relationships and pressures on TVNZ Ltd since it was established as an SOE. These relationships and pressures are diagrammed in figure 7.1. The effects of deregulation and the entry of competitors was discussed at length in earlier chapters. In this chapter, we discuss the relationship between the board and management, between the company and the shareholding ministers and their advisers, and between the company and the Broadcasting Commission. The final subsection considers political and social pressures on the organisation.

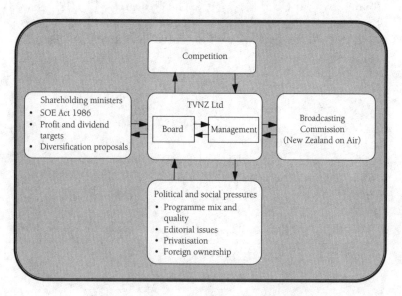

FIGURE 7.1: RELATIONSHIPS AND PRESSURES

The Board and Management

As discussed in chapter 4, members of the initial board of TVNZ Ltd were not invited to join the board until after the valuation itself was completed and agreed to by the shareholding ministers. Acceptance of their appointment to the board constituted an acceptance of this asset valuation and financial structure. Unlike other SOE establishments, board members were given no part in negotiating a sale and purchase agreement. The only board member who had been privy to the underlying discussions on valuation was Brian Corban, the chairman-designate of TVNZ Ltd, by virtue of his membership of the Ministerial Advisory Committee.

Moreover, it will be recalled that during the short period in which the Ministerial Advisory Committee was operating, a decision vacuum resulted because TVNZ's management was caught between a caretaker BCNZ board and a Ministerial Advisory Committee which lacked the executive authority of an establishment board. Consequently, at the first meeting of the new board on 8 December 1988, members were confronted with a chief executive leading a team of managers who wanted to forge ahead quickly with major initiatives to restructure the company. They also faced a chairman who firmly supported the chief executive and was asking the board to take the risk of going along before they had a chance to become totally familiar with the company and its business. Although extremely uncomfortable with what they were being asked to sanction at this first meeting, the board endorsed significant restructuring plans involving the reorganisation of drama production, setting up Avalon and Outside Broadcasts as separate subsidiaries, rationalisation of the design department, staff reductions, wide-ranging delegations to the chief executive for personnel and capital expenditure decisions, a joint venture with Lion Nathan, development of a plan to invest in Sky Network Television and TVNZ's business plan for the first six months of operations.

Typically, the new board can be expected to provide the initial challenge to the status quo where an organisation is established as an SOE. However, this was not the case at TVNZ. The challenge did not come from the board, at least not at first. Partly as a consequence of the design of the process it came from the chief executive and the senior management of the company. With the exception of the chairman and one other board member who had also served on the last

BCNZ board, board members were caught on the back foot. Struggling to catch up with the state of play they could act only as the overseers rather than the instigators of change. Without the initial challenge that Julian Mounter made to the board above him and to his managers beneath him, the outcome for TVNZ Ltd might have been very different.

During his time as deputy chairman of the BCNZ, Brian Corban had become frustrated with its bureaucratic structure and its lack of a coherent strategy. Interestingly, Corban had been a member of a subcommittee of the BCNZ board which was responsible for hiring Julian Mounter as director-general of TVNZ. Corban recalls that at that time he was convinced that the problems in the BCNZ were such that they needed a person with wide international experience in television. In Brian Corban, therefore, Julian Mounter found an ally who was willing to accept the personal risk involved in supporting his chief executive's strategic vision and his drive to change the organisation before it was overwhelmed by new competition. On this Corban remarked:

> It's easy now to look back and say well that was successful and take it all for granted. But at that time there was no guarantee of success. Julian summed it up in October 1988 when I was announced as chairman designate. He said: 'Brian we've got a 10% chance, a 10% window of opportunity'. I said 'Fine, we'll go for it and what our job is is to keep expanding windows. We've got to turn it around, it's got to become 90%.'

As chairman, Corban gave a great deal of time to company affairs, a matter which caused concern among Treasury officials who believed that the SOE chairpersons should confine themselves primarily to a monitoring and oversight role. On this one board member provided the following perspective:

> The bureaucrats never understood that this company had to relate to other [broadcast media] companies. We were unusual in the sense that it was the chairman of TVNZ Ltd who was expected to call on the chairman of the BBC, rather than the chief executive alone. So they went together because Julian Mounter put the pressure on Brian to do that and he had to do it as part of the job. It was part of the culture.

In the development of TVNZ Ltd Corban appears to have played two roles. First, he attempted to provide the chief executive and his senior management team with 'space' to restructure the organisation by creating a buffer between ministers and their advisers and the company's management.

> I stood between the company and the politicians. With a broadcasting SOE I think it's extremely important that the chairman has an educated sense of constitutionality. Integrity and independence for a broadcasting organisation and its chief executive, who is also the editor-in-chief, is important. The business of TVNZ is invading the homes of virtually every New Zealander on a daily basis. The stake of politicians in how they are being portrayed is very high so the role of chairman is onerous and the function of the chairman has some peculiarities that aren't relevant in other SOEs.

The chairman also played a role in questioning, challenging and providing a sounding board for the chief executive on the direction in which he was attempting to take the company. It was not long before the two men spoke to each other almost every day as Mounter came to value regular discussions with the chairman about the myriad of strategic, political, and social pressures that confronted the company on a day-by-day basis. On this Mounter commented:

> Brian never interfered in the running of the company but he had a very good way of making me look at things again by asking questions: 'How much work have you done on this, how does it fit, are you really sure you've made the right decision?'. He would test you very well on it and to that extent he was an enormous help.

The sequence often involved the chief executive persuading the chairman of the need for a particular decision. The chairman would then assist the chief executive to persuade the board. One board member, Tipene O'Regan, believes that the chairman and the chief executive complemented each other well. Corban, he commented, was 'a reflective man who provided a good foil for the sharp, fast moving and analytic Mounter'.

While both of these men had a strong commercial orientation, each also professed a philosophical commitment to public service standards of television programming, a commitment which was challenged by many cultural critics of TVNZ's new direction. This point was recognised by Corban himself who commented:

> Julian and I never let go of the public service qualitative standard. Many observers would say we did but I can tell you personally that we never did. The problem was that we had to use every ounce of commercial

ability we had to establish a viable future for the company to enable us to get back to that sort of a vision. It's a very tough standard and Julian and I both believed that we should try to bridge the gap.

While cultural critics may have thought TVNZ's first board was overly commercial, Treasury officials and shareholding ministers had concerns that it was not commercial enough. This led to changes aimed at strengthening the commercial orientation of the board. As discussed in chapter 4 the board was expanded in late 1989 and further changes were made when the terms of three board members expired at the end of 1990. The objective of these changes was to strengthen the board's commercial orientation.

TVNZ's Board and Shareholding Ministers

In the first two years as an SOE, the relationship between TVNZ's board and the shareholding ministers and their advisers was a difficult one. The major issues generating conflict and dissension in this relationship involved disagreements over (1) the principal objectives of the company, (2) profitability targets and dividend levels, and (3) the extent to which TVNZ should be allowed to diversify and expand its business activities. After discussing each of these issues we provide an assessment of the reasons which appear to lie behind the relationships as they developed.

Disputes over TVNZ's Objectives

As discussed in chapter 2, the Officials' Co-ordinating Committee on Broadcasting had recommended that the new broadcasting SOEs, and TVNZ Ltd in particular, should have an obligation to encourage locally made programmes that reflect and develop New Zealand's identity and culture. To that end, the Officials' Committee recommended that conditions be placed in TVNZ's Statement of Corporate Intent (SCI) to ensure that the corporation accepted this obligation as its predominant objective. This recommendation, among others, was accepted by Cabinet on 22 August 1988, and board members, who were appointed by the Minister of Broadcasting, received letters of appointment containing the following instruction:

Television New Zealand Ltd will be required to fulfil its obligation to be a successful business pursuant to Section 4 of the State-Owned Enterprises Act 1986. Within that context Television New Zealand is to have the *predominant* objective of reflecting New Zealand's identity and culture and to encourage New Zealand programming. This objective is to be contained in Television New Zealand's Statement of Corporate Intent. [emphasis added]

Treasury officials, whose responsibility it was to advise shareholding ministers, objected to this wording, arguing that it conflicted with Section 4 (1) of the State-Owned Enterprises Act 1986 which states unequivocally that 'the principal objective of every state-owned enterprise shall be to operate as a successful business'.[1] On this matter, however, Treasury officials did not carry the day and were overruled by a senior official of the Ministry of Commerce. The letters of appointment with the disputed wording were sent to board members.

As a result, a situation was created where there was disagreement about the new objectives of the company between different ministries and more than a little confusion on the part of board members. Indeed, the interpretation the board placed on the letters of appointment reflected the composition of TVNZ's initial board. More commercially oriented board members could well have interpreted the wording in the letters differently. It was not surprising, therefore, that the first SCI submitted by TVNZ Ltd's board was unacceptable to Treasury officials because it appeared to place the achievement of commercial objectives secondary to the achievement of social objectives. Treasury officials quickly advised shareholding ministers that the board's policy, if carried out, would

> . . . reduce profits, reduce revenues to the Crown, reduce asset value and thus contravene the State-Owned Enterprises Act. Even more fundamentally, this policy would so confuse the Company's goals that it would be impossible to assess its performance objectively—effectively a return to the situation which prevailed under the former Broadcasting Corporation of New Zealand when there was confusion between social and commercial objectives and continuing arguments about financial structure and performance.

A review of subsequent communications between Treasury officials, shareholding ministers and company officials showed quite clearly that the board had been placed on the horns of dilemma and had great difficulty in reconciling the requirement to act commercially placed on them by the SOE Act with the social, public service broadcasting objective required by their letters of appointment. A further problem was caused in this respect by uncertainty about the level of funding TVNZ would receive from the Broadcasting Commission for 'non-commercial' programming. Therefore, in mid 1989, the board turned to the shareholding ministers to seek assurance that TVNZ would receive sufficient funding to continue to provide the level of programming reflecting New Zealand identity and culture that the board considered necessary to meet its social objectives. Shareholding ministers refused to provide this assurance because, in their view, it would violate the separation of commercial and social activities which was the focus of the new broadcasting regime and the reason that the Broadcasting Commission had been established. Following a series of letters and other communications on this issue matters came to a head on 12 July 1989 when the board sent ministers a revised draft of the SCI containing the following statement:

> The company will strive to achieve certain social objectives in the production and broadcasting of television programmes. It will seek funding for the achievement of these social objectives from the Broadcasting Commission and where the cost is not provided by the Broadcasting Commission it will seek funding of these from the Crown in accordance with Section 7 of the State-Owned Enterprises Act. If full funding is not provided, TVNZ Ltd will cross subsidise this activity from commercial activities to the extent that it considers desirable.

This quickly drew a sharp response from the shareholding ministers, who replied on 21 July:

> We believe that the Board would contravene the Act if it took over this responsibility from the Commission as is implied by the Board's policy to support non-commercial programmes 'to the extent it thinks desirable'. If the Commission chooses not to support certain non-commercial programmes proposed by TVNZ and if no other outside funding is available, then the Board must decide, within the framework set out in the State-Owned Enterprises Act, what its commercial response should be.

On 28 July, the board chairman again wrote to the shareholding ministers stating that the board was unwilling to conclude an SCI on terms other than those already provided. In this letter the board argued for (1) a different interpretation of the SOE Act which would permit the company to use commercial revenue for the preservation of non-commercial programmes, (2) an increase in government funding available to the Broadcasting Commission or additional funding under the section 7 of the State-Owned Enterprises Act to bridge the gap between the $20–$25m TVNZ expected to receive from the

Broadcasting Commission and the $36m TVNZ estimated it currently spent on 'non-commercial' programming, or (3) an instruction from shareholding ministers to remove public service programming from their schedule. Shareholding ministers, on the advice of Treasury officials, who at this point were starting to wonder if they had made any progress with TVNZ's board on this point, rejected all these options, pointing out once again to the board that it was the Broadcasting Commission which was accountable for decisions to support non-commercial programmes proposed by TVNZ. This, the ministers stated clearly, was 'not the responsibility of the board or the Shareholding Ministers'.

Now under considerable pressure from shareholding ministers and their Treasury advisers to urgently reconsider their position, the board continued to resist but finally gave way and revised the disputed statement in their SCI to read as follows:

> The Company will strive to achieve certain social objectives in the production and broadcasting of television programmes. It will seek funding for the achievement of these objectives from the Broadcasting Commission. This Statement of Corporate Intent has been prepared on the basis of receiving $36m per annum from the Broadcasting Commission to assist in maintaining current levels of public service programming and non-commercial activity.

This statement was considered to be consistent with the requirements of the State-Owned Enterprises Act 1986 and therefore acceptable to shareholding ministers. Instead of continuing to lobby shareholding ministers for increased funding for non-commercial programming, TVNZ officials instead pressed the Minister of Broadcasting to increase the level of funding to the Broadcasting Commission. In the view of Treasury officials this approach was acceptable because it recognised the separation of functions inherent in the new structure of broadcasting institutions.

The confusion over objectives in this case stemmed from the initial attempt to require TVNZ to meet social as well as commercial objectives. This created a quandary for board members and led to contentious debate over what should be the primary objectives of the new SOE. The board were confused about their role and subject to conflicting pressures from different government ministers and from officials. As Brian Corban, the chairman of TVNZ through this period, recalls:

The letters of appointment of the first board said one thing but Treasury intended another thing. What the Minister of Broadcasting intended was that there be an equal balance between the commercial and the cultural. What Treasury intended was that the commercial objective be paramount and the cultural and social objectives be secondary or subservient. And that actually led to a problem for board members in the definition of their responsibilities and led to a major problem in the negotiation of the SCI and some quite heated discussions with Treasury officials.

Treasury officials, however, argue that it was not a matter of Treasury intent. In their view, the board's interpretation of their letters of appointment was simply *ultra vires* the State-Owned Enterprises Act. For them, the SOE Act required the separation of functions and that was the bottom line as it was their responsibility to ensure that the Act was followed.

Disagreements over Profit and Dividend Targets

While the wording of the company's objectives was now acceptable to shareholding ministers, allowing TVNZ's first SCI to be tabled in Parliament, Treasury officials remained concerned that the low dividend payout proposed by the board (10 percent of after-tax profits) was well below that which was normal for other SOEs, other New Zealand companies and even other overseas broadcasting companies. Treasury officials advised shareholding ministers that acceptance of such a low dividend payout would give the appearance of undermining the Broadcasting Act by permitting the continued support of non-commercial programmes with revenues from commercial programmes, grant a competitive advantage to TVNZ by allowing it maintain a large supply of cash for its battle with TV3, reduce the pressure on TVNZ to become more efficient and lower its costs and establish an undesirable precedent for dividend payouts by other SOEs. Instead, Treasury proposed a minimum dividend payout of 40 percent of after-tax profits. Ministers wrote to the board along these lines a number of times in the ensuing months but the board tenaciously defended its position arguing that it could not recommend 'a dividend rate unrelated to the business realities manifest in the company's business plan'.

With the board's presentation of proposed rate-of-return targets and dividend payout levels for its 1990 SCI and business plan, the dispute over the dividend payout spread to include profitability targets. TVNZ initially targeted a 6 percent rate-of-return target on shareholders' funds for 1990 coupled with a continuation of a 10 percent dividend payout ratio. Neither of these targets was acceptable to shareholding ministers who began to place considerable pressure on the board to raise not only their profitability and dividend payout targets for 1990 and the two subsequent years, but also their actual dividend payout for 1989. Treasury advised ministers that TVNZ's after-tax rate of return on shareholders' funds for 1989 was expected to be around 16.2 percent. In their view the company should be proposing a target rate of return on shareholders' funds of around 15–16 percent for 1990 and a dividend payout of 40 percent of net profits after tax.

Following a series of increasingly heated meetings and pointed communications between the shareholding ministers and the board chairman, the board was persuaded to raise its target rate of return on shareholders' funds for 1990 from 6 percent to 7.2 percent[2] and then after additional pressure (and a resulting decision by the board to downsize and restructure the business during the year) to 9.8 percent. Thereafter the board projected that the after-tax rate of return on shareholders' funds would rise to 13.9 percent in 1991 and 16.5 percent in 1992. (See table 8.2 in the next chapter.)

TVNZ Ltd's recommended targets for 1990 and 1991 were considered to be well below commercial norms by Treasury officials, who advised the shareholding ministers that the low rates proposed raised serious questions about the board's commercial orientation and decision making processes. Because they were also worried about the risk inherent in TVNZ's plans for overseas investment and expansion against a backdrop of incomplete plans and inconsistent financial reporting by the company,[3] Treasury officials took the unusual step of recommending that the shareholding ministers invoke section 18 of the State-Owned Enterprises Act 1986 to gain additional information about the board's decision making processes.[4] On 28 May 1990, therefore, shareholding ministers used this authority to request from the company

> . . . copies of all reports or other information supplied to the Board on which the Board relied in reaching a decision that a 9.8% rate of return was consistent with the requirements of the Act . . . [and] copies of those portions of the minutes of the Board meetings dealing with the rate of return.

While a review of these reports and board minutes by Treasury analysts and members of the SOE Advisory Committee proved inconclusive, officials remained concerned about the extent to which TVNZ's board was challenging management to cut costs and to pursue options to improve short-term profitability. In actuality, board papers indicate that actions were being taken in the company to cut costs at this time; the difficulty was that these cuts were not expected to result in major savings until 1991.

As it turned out, the 9.8 percent after-tax rate of return was too much for the company to achieve in 1990. A weaker-than-expected overall advertising market, the loss of approximately $40m of advertising revenue to TV3, reduced funding from the Broadcasting Commission and a slow start to TVNZ's downsizing and restructuring programme combined to force the board in October 1990 to revise its forecasted profitability downwards to 6.8 percent for 1990, 10.6 percent for 1991 and 15.1 percent for 1992. As a material portion of even this 6.8 percent return was generated by a change in accounting policy,[5] the deterioration in profitability was even greater than Treasury officials had feared. Officials immediately wrote to shareholding ministers expressing their concern and once again questioning TVNZ's strategy of international expansion when it was unable to earn an 'acceptable' level of profits on its core business activities. Shareholding ministers subsequently wrote to TVNZ pointing out that they were by nature 'risk averse and unwilling to supply additional capital, particularly . . . in light of the government's fiscal situation'. Facing the prospect of lower than expected profitability for 1990, the ministers chose to reinforce a point made by the company's deputy chairman 'that the outcome for 1991 will represent a true test of the Board's performance and would justifiably be the basis on which the shareholders should judge the future of the Board'.

On the dividend issue, the board finally acquiesced to persistent pressure from the shareholding ministers to increase their dividend payout for 1989 and to change their dividend payout policy. The board ultimately paid out $8m of actual after-tax profits for 1989 and, after even more debate, rewrote the company's dividend policy to meet the ministers' repeated requests for a dividend payout of 40 percent of net profits after tax and extraordinary items. TVNZ performance against its SCI targets is further discussed in chapter 8.

Debates over Diversification Proposals
As discussed in chapter 5, an important part of TVNZ Ltd's strategy

was to look for expansion opportunities that would complement either its core broadcasting or its signal distribution and transmission businesses. Consequently, in its first SCI submitted on 12 July 1989, the board proposed that its scope of activities extend beyond broadcasting to include 'the production of telecommunications services both within New Zealand and overseas'. However, shareholding ministers opposed such an expansion of scope until such time as BCL had been valued, its financial structure determined, and the board had prepared and submitted a detailed and specific plan for any new business ventures in telecommunications for review.

Two months later, on 18 September, the chairman notified ministers that the company had entered an agreement with Bell Canada to prepare a plan for establishing an alternative telephone company and wished to publicly announce that agreement. Once again, ministers wrote to TVNZ's board asking it to wait, this time for the government's soon-to-be-announced policy on SOE diversification. However, ministers did note that they were less than enthusiastic about the proposed expansion, particularly in light of the board's proposed 10 percent dividend payout and its low profitability targets. Ministers pointed out that, in general, their preference was 'that funds which should be returned to shareholders as dividends should not be spent on new business ventures'.

On 17 October 1990, the shareholding ministers wrote to all SOEs setting out the government's policy on diversification. Under this new policy SOEs were informed that, in general, they

... should focus on maximising the value of their core business, constrained by the shareholders' preference that the business should not expand beyond activities related to the core business. Within this constraint investment decisions remain the Boards' responsibility.[6]

The question now was whether the diversification into telecommunications proposed by TVNZ and its subsidiary BCL should be approved under this new policy. As the proposed investment was outside the core activities of TVNZ as stated in their SCI, the test under the new policy was whether the investment was expected to 'add value to the core business'.

Treasury officials concluded that TVNZ's proposed investment in the alternative telephone company appeared to meet this test and therefore should not be prohibited. Nonetheless, they continued to worry about the risk associated with the investment and advised ministers that, before approving an extension to the scope of TVNZ's business activities, they should be satisfied that TVNZ's board had explored and analysed alternatives to a straight equity investment. Having received additional information from the company that provided this assurance, ministers wrote to the board on 3 May 1990 stating that if the board concluded that the best way to maximise the value of the core business was to expand into telecommunications, then they would approve an appropriate modification to TVNZ's SCI. This cleared the way for TVNZ to obtain 25 percent of the equity of Clear Communications Limited.

Much more controversial than the investment in the alternative telephone company was an ambitious plan to buy an equity share in Australia's Channel Nine network, which was owned by Bond Media Limited. Referred to in the company's business plans as part of a wider strategy to protect product supply, obtain distribution agreements, and enhance its overall strategic position, this scheme called for the company to seek strategic investments in offshore broadcasting, production, and distribution companies. To finance these strategic investments, TVNZ's intention was to sell its surplus land and buildings. The company also planned to sell up to 49 percent of the equity in its subsidiaries (SPP, BCL, and Avalon)[7] to organisations, such as offshore television companies, which could, in turn, add value to the TVNZ group by placing business with these subsidiaries.

It was the view of TVNZ's management that the restructuring of the Australian television industry taking place at that time provided them with a unique opportunity to take an equity share in a major Australian television broadcaster at a reasonably modest price. In February, therefore, the board notified shareholding ministers that it had made a preliminary offer for 10 percent of Bond Media Limited shares subject to the approval of the shareholding ministers and ratification of a variety of other conditions including a 'due diligence' assessment. The company also informed ministers that in addition to an equity investment, it was negotiating a number of subsidiary arrangements with the owners of Channel Nine that would enhance the profitability and strategic position of TVNZ's commercial television business. These included the savings from sharing of offices and satellite facilities, non-compete clauses in the Pacific where joint ventures would be possible, programming arrangements and the sale of a share of South Pacific Pictures to Bond Media Limited.[8]

Although the proposed investment fell within TVNZ's approved scope of business activities, officials at the Treasury and in the office of the Minister for State-Owned Enterprises reacted negatively to TVNZ's proposal, calling into question the commercial wisdom of such an arrangement because they believed it might expose the Crown to unacceptable levels of risk. Given the low rates of profitability TVNZ was projecting for 1990, Treasury officials advised ministers that board and management attention should be focused on improving the profitability of the core broadcasting business rather than on the investigation and negotiation of risky offshore investments. TVNZ's response was that investments in both Bond Media (and the alternative telephone company) were essential to the long-term profitability of the company's core television business, and suggestions to the contrary simply reflected 'a lack of business expertise'.

In the end it was the Maori Council proceedings against the broadcasting assets which stopped TVNZ Ltd from continuing to consider its intended purchase. Had TVNZ owned its assets at this time, the company might have been in a position to raise the necessary finance by selling surplus assets and shares in subsidiary companies. However, because the licence agreement under which the assets were operated by TVNZ required shareholders' approval for not only the sale of surplus assets and shares in subsidiaries but also to increase the company's borrowing limits, the shareholding ministers were able to stop the company from proceeding with the investment. Acting on advice from officials that TVNZ had failed to make a strong enough case that 'the strategic benefits justify the substantial risks associated with the minority investment' in Bond Media, shareholding ministers wrote to TVNZ on 3 May withholding the approvals sought by the company to effect a sale of shares in subsidiaries[9] or to increase its borrowing limits. Without these approvals the company was not in a position to raise the $32.5m necessary to finance the investment in Bond Media Limited or implement its wider strategy of selling shares in its subsidiaries to organisations which could add value to the TVNZ group by placing work with them.[10]

Factors Underlying TVNZ's Relationships with its Shareholding Ministers

There appear to be a number of underlying reasons for the contentious relationship that developed between TVNZ's board and management and the shareholding ministers and their advisers during the transition period. What influenced this relationship most appears to have been the manner of the first board's appointment which led to a protracted debate with the board over their obligations under the SOE Act. This experience, combined with problems with the quality of TVNZ's financial reporting at this time, materially affected the views of Treasury officials, who concluded that some of the problems shareholding ministers were experiencing in their relationship with the company probably stemmed from a lack of a high level of commercial expertise and experience on the part of some board members and an inability to come to terms with the new environment.

Ironically, on the other side, board members and senior company executives expressed their frustration at having to deal with Treasury officials and shareholding ministers who they believed lacked an adequate understanding of the commercial challenges which faced TVNZ Ltd in its newly deregulated environment. One board member commented:

> Unless its shareholders wake up this company is in serious trouble. We have the appalling problem of a shareholder who does not understand the asset. We are probably the most successful company in the world of our kind but we have a shareholder who is not prepared to make an investment, is risk averse and is not wanting, for political reasons, to let anyone else have a slice of the company.

What appears to have most frustrated TVNZ's board and management was that the Treasury officials charged with monitoring their performance seemed to be giving insufficient weight to the recommendations of the Rennie Committee,[11] the work of the Ministerial Advisory Committee and specifically, the assumptions which underlay the valuation of TVNZ's assets. In their report delivered in November 1988, the valuers, Fay Richwhite, assumed in setting up the valuation that the company would follow certain strategies which would affect the returns from the core business in the first three years of competition. In particular, the valuers assumed that TVNZ management would be permitted and encouraged to restructure production activities so as to reduce expenditures, to sell shares in subsidiaries so as to achieve improved utilisation of facilities, to invest on a fully commercial basis in overseas enterprises (where this was required to ensure access to programme sources and export outlets directly related to the defence and improvement of TVNZ's core businesses), and to develop new forms of television services that would compete against and displace its core businesses.

Realising that it would take time for the benefits from these policies to start to flow, Fay Richwhite recognised that there would be a substantial drop in returns for the core TVNZ business in the first three years of competition. Indeed, the valuers estimated the return on equity (expressed in 1988/89 dollars) to be 16.1 percent, 7.2 percent, 2.4 percent and 3.2 percent for the years from 1989 to 1992 respectively. It was not until 1995 that the return was expected to be greater than 10 percent.

It was on this basis that the deputy chairman, Rob Challinor, who had initially advised shareholding ministers on the valuation, took issue with Treasury's advice that TVNZ should be projecting a target rate of return of around 16 percent in 1990 and subsequent years. In a memorandum dated 30 May 1990 to TVNZ Ltd's chairman and chief executive (a copy of which was sent on to the shareholding ministers) Challinor questioned Treasury's implicit assumption that TVNZ's valuation had been set in a manner that made a 16 percent return reasonably achievable in the second year of operation. To quote from the memorandum:

> The equity of TVNZ Ltd was set at $140m. This was considered to be fair value based on the potential returns of the company over a period of time given its strategic position and potential in the television industry. The equity value was not set based on the profits and returns which could be achieved in the short run. To the contrary it was acknowledged that profitability would fall during the initial period of competition and that good profits would only be sustainable following the adoption of the strategies [referred to in the valuation document]. The monitoring of the performance of TVNZ Ltd based on an initial equity of $140m must take this factor into account.

Challinor went on to argue that TVNZ was performing better than it was receiving credit for. He contended that a number of things needed to be considered when monitoring the company's performance. The first was the fundamental change in the competitive position of TVNZ, which had faced almost complete deregulation of the broadcasting industry. The second was that the valuation was set independently and was not the subject of a prolonged negotiation between the new directors and the government as had been the case with other SOEs. Had such negotiations been allowed in the TVNZ case, a lower valuation might have resulted and the sought-after rates of return on shareholders' funds made achievable. Third, it was inappropriate to directly compare TVNZ's rate of return on shareholders' funds with the average rate of return on shareholders' funds for other New Zealand companies because these returns were, for the most part, based on historical cost methods of asset valuation. Because TVNZ's assets were revalued as at 30 November 1988, it would inevitably suffer in such a comparison. Lastly, TVNZ's performance should be considered in the light of the recent poor performance of media companies around the world.[12]

To determine if TVNZ was actually performing better than had been predicted in the valuation, Treasury sought independent advice in the middle of 1990 from a major firm of public accountants. From their review the consultants were unable to say whether the performance of the group to the date of the review was due more to good luck than good management. However, the point remains that up until the time of the review, neither the recommendations of the Rennie Committee, nor the assumptions underlying the valuation, were generally referred to in Treasury analyses of the targets proposed by the company, the proposals put forward by the company to move into telecommunications, or proposals to invest overseas. Instead, Treasury analysts focused their reports to ministers almost exclusively on the implications of proposed actions on the short-term profitability, dividend payouts, and risk profile of the company.

The original differences of opinion over objectives were heightened by this difference in time horizons. Whereas TVNZ's management focused on positioning the company competitively and putting into place the strategies they believed were necessary to increase future (albeit uncertain) cash flows and returns, shareholding ministers and their Treasury advisers focused on pressuring the company to improve short-term financial performance and dividend payouts and to contain the Crown's risk exposure. This focus on certain profits and dividends today, rather than on an increase in value in anticipation of uncertain future growth, appeared to have underlain much of Treasury's questioning of the company's diversification plans. Part of the problem from the perspective of Treasury officials was that they experienced great difficulty in getting financial information on diversification proposals from the company.

However, some of these difficulties went beyond differences of opinion and involved communication problems between the company and the shareholding ministers and their advisers. In some cases, in the race to change the company internally and to diversify, the company did not always clearly communicate the details of its strategies, plans, and operating results to its shareholders. Julian Mounter has since acknowledged that this was the case:

> When Cliff Lyon came onto the board he explained to us that we were missing the point on how to explain ourselves. I thought we were telling them [shareholding ministers and Treasury] what they wanted to know and we arrogantly assumed that they would understand. What we were doing was communicating with our shareholders and we should have realised that a bit better.

The fact that Treasury analysts also found it difficult to compare actual against planned results because of repeated reforecasting of business plans and changes in accounting methods (the effects of which were not built into plans) caused Treasury officials to question company figures. This suspicion did not help the development of a good relationship between the company and shareholding ministers and their advisers during the transition period. A partial cause of this problem was the poor state of TVNZ Ltd's financial systems in the 1989 calendar year.

At least from the board's perspective, communications were hampered by the location of TVNZ's headquarters in Auckland rather than Wellington and by the need over the 1988–1992 period to deal with seven different shareholding ministers and two Ministers of Broadcasting and their advisers. In addition, as mentioned in chapter 7, the board also had to deal with political confusion as to the different roles of the Minister of Broadcasting and the shareholding ministers.[13]

TVNZ and the Broadcasting Commission (New Zealand on Air)

The separation of commercial and social objectives in broadcasting and the establishment of new organisations which would pursue each of these objectives independently, lay at the heart of the new reforms. The pursuit of commercial objectives was to be the role of the broadcasting SOEs, whereas the pursuit of social objectives was to be the function of the Broadcasting Commission, which was to act as a specialist purchasing agency. Its primary source of funding would be the public broadcasting fee.

This arrangement had three underlying objectives. The first was to establish a mechanism by which the Minister of Broadcasting and

Cabinet could be assured that public service objectives would continue to be pursued in a deregulated, commercially focused broadcasting environment. As the commission would be subject to ministerial direction the government would retain the ability to influence the social objectives it wished to emphasise. The second and related reason was to improve accountability for the public broadcasting fee, which under the old regime had been collected directly by the BCNZ. Few details of how this considerable sum of money had been spent appeared in the annual reports of the BCNZ. The third objective was to encourage independent producers who would compete with the television networks for programme funding.

The Officials' Committee recommended that the primary objective of the Broadcasting Commission 'should be to achieve adequate levels of universal coverage, local content and minority interest programming' with a great deal of emphasis being placed on achieving sufficient local content to reflect and develop the New Zealand identity and culture. This emphasis is reflected in the wording of the Broadcasting Act 1989 which in section 36 sets out the functions of the commission:

> The functions of the Commission are—
> (a) To reflect and develop New Zealand identity and culture by—
> (i) Promoting programmes about New Zealand and New Zealand interests; and
> (ii) Promoting Maori language and Maori culture; and
> (b) To maintain and, where the Commission considers that it is appropriate, extend the coverage of television and sound radio broadcasting to New Zealand communities that would otherwise not receive a commercially viable signal; and
> (c) To ensure that a range of broadcasts is available to provide for the interests of—
> (i) Women; and
> (ii) Children; and
> (iii) Minorities in the community including ethnic minorities; and
> (d) To encourage the establishment and operation of archives of programmes that are likely to be of historical interest in New Zealand— by making funds available, on such terms and conditions as the Commission thinks fit, for—
> (e) Broadcasting; and
> (f) The production of programmes to be broadcast; and
> (g) The archiving of programmes.

Sections 37 and 39 of the Act set out how the commission is to implement its functions. Section 37 is entirely devoted to the promotion of New Zealand content in programming while section 39 is devoted to setting out those matters to be taken into account in assessing funding proposals. These include the extent to which funding applicants have access to other funds, the potential size of the audience, the extent to which the proposed programme would contribute to meeting the commission's objectives under section 36 (a) and (c), the availability of a balanced range of programmes for varied interests and the likelihood that the proposed programme, if produced, would be broadcast.

It is important to note that the words 'public service' or 'non-commercial' do not appear in the Broadcasting Act, probably because they are legally difficult to define. Instead, what the Act does is to translate the notion of 'public service' broadcasting into the notion of New Zealand content, making the latter a major focus of the Act. At the same time no distinction is drawn between 'commercial' and 'non-commercial' programming. The Act is also silent on the issue of 'quality' in programming. It was left to the Minister of Broadcasting to raise the issue of quality programming when he briefed the commission on its responsibilities under the new Broadcasting Act.[14]

It was against this background that the commission set to work to define its goals. For the period 1989 to 1992 five goals were agreed and set out in the commission's first annual report for the year ended 30 June 1990:

1 To ensure that mainstream audiences have access to a variety of quality programmes made for New Zealanders, by New Zealanders and about New Zealanders.
2 To facilitate diversity in broadcasting by supporting a range of broadcasting opportunities for all audiences and programme makers, including minority and mainstream interests.
3 To respond to public opinion and to monitor, react to and influence the broadcasting environment.
4 To support Maori broadcasting aspirations.
5 To maximise fee collection and achieve cost-effective results.

Because they were reliant on the public broadcasting fee for their funding, the commission set out to establish a public persona. They did this by naming the product of the commission 'New Zealand on Air'. The first executive director of the commission, Dr Ruth Harley, commented:

We had a serious identity problem. The public confused us with the BCNZ. We also had the problem of trying to extract money from the public to fund an organization called the Broadcasting Commission which had no clear purpose in the public's mind. So we tried to give ourselves a brand which identified the public's money with the result. We initially thought of things like 'The Voice of New Zealand' but that sounded a bit fascist and so we ended up with 'New Zealand on Air'.

Sorting Out the Relationship

It was not long before the commission, now known as New Zealand on Air, was at loggerheads with TVNZ's board and executives over a wide range of issues centred on New Zealand on Air's interpretation of its role under the Broadcasting Act and the types of programmes it started to fund. Julian Mounter recalled:

> When it [the Broadcasting Commission] was announced I thought 'Ah, so we get on with the commercial bit and ballet, Maori programmes, specialist drama—everything that the commercial broadcaster isn't going to make, will be funded by the Commission'. We got some money out of them [for these types of programmes] but we then saw them putting money into [TV3] programmes which were highly commercial like *The Billy T. James Show*, *Black Beauty* and *Sixty Minutes* which were designed to hit our ratings.

TVNZ's directors and executives argued that it was the role of commercial television to provide these types of programmes and that New Zealand on Air's funding of them constituted an intervention in the commercial market-place. TVNZ argued that New Zealand on Air ran the risk of distorting normal commercial competition by effectively subsidising ratings-oriented commercial programmes on New Zealand's three commercial channels. More importantly, if New Zealand on Air continued to fund commercial programmes, there would be less money available for non-commercial programming. Mounter worried that, if they were forced to drop regional news or such things as ballet through lack of funding, then eventually the public and the ministers would say 'right, enough of this, we'll change it all again'. Brian Corban put it as follows:

> We believed that the proper interpretation of the Act was that the Commission should fund non-commercial broadcasting, that it should not fund commercial programmes that a broadcaster would justify under commercial principles. Under the SOE Act, TVNZ has very restricted

room to apply our funding to non-commercial programmes. But if we and TV3 funded our own commercial programming this would leave more money in the pot, for not just TVNZ [and TV3] but also independent producers to produce non-commercial, minority and special interest programmes that wouldn't otherwise be produced or shown.

However, this interpretation was never the intent of those officials who had helped to draft the Broadcasting Act 1989 or of Jonathan Hunt, the Minister of Broadcasting at the time.[15] Jim Stevenson, who chaired the Officials' Co-ordinating Committee on Broadcasting and who later sat on the Broadcasting Commission, commented:

New Zealand on Air helped fund *Shortland Street* and *Marlin Bay* on TVNZ and *Homeward Bound* on TV3. These all have a strong commercial component. New Zealand on Air's objective is not to fund non-commercial television programmes but rather to promote New Zealand content. New Zealand on Air promotes this by providing seed funding or supplementary funding to get the programme on air.

Treasury officials shared this view from the beginning, writing to the shareholding ministers on 7 December 1989 that 'TVNZ appears to be of the view that the Broadcasting Commission cannot fund even in part what TVNZ considers to be commercial programmes. This is not our reading of the Broadcasting Act.'

The philosophical clashes between TVNZ and the commission arose partly because the State-Owned Enterprises Act 1986, which controls the activities of TVNZ, and the Broadcasting Act 1989, which controls the activities of the commission, could be interpreted as leaving a gap between them. As noted in the prior section, the argument has been made that the State-Owned Enterprises Act specifically precludes TVNZ from engaging in 'non-commercial' behaviour.[16] It is this which appears to have been the basis for TVNZ's position that the commission should focus its funds on non-commercial programming, i.e., minority-interest and special-interest programming such as Maori language programmes, religious programmes, and ballet.

The Broadcasting Act, on the other hand, does not require the Broadcasting Commission to focus its resources only on non-commercial (i.e., minority-interest and special-interest) programming. Rather, the Act requires the commission to promote a range of programming catering to a variety of interests and, in particular, to promote programmes which reflect and develop New Zealand's identity and culture. New Zealand on Air argued that to confine itself solely to funding only non-commercial (minority-interest and special-interest) programming would mean they could not finance other types of more popular programming that promote New Zealand's identity and culture and which would not otherwise be produced for New Zealand television. Ruth Harley pointed to New Zealand on Air's subsidisation of the New Zealand 'soap' *Shortland Street* as an example. Even though *Shortland Street* became a major success in the ratings, New Zealand on Air's position was that it is unlikely that the programme would have ever been made without a contribution from New Zealand on Air. TVNZ invested $7.5m in the production of the first set of *Shortland Street* episodes, with New Zealand on Air contributing $2.5m.

Conflict over the role of New Zealand on Air in commercial television also resulted in day-to-day jousting over a variety of operational details, including the methods of contracting and programme scheduling, the use of 'New Zealand on Air' as the name of the Broadcasting Commission, on-screen crediting arrangements for programmes assisted by New Zealand on Air, advertising and promotion of New Zealand on Air and assisted programmes, the role of New Zealand on Air as an investor in programmes, and TVNZ's requirement that its facilities should be used for the production of all programmes with which it is associated.

In terms of contracting, TVNZ preferred that New Zealand on Air bulk-fund the production of a range of programmes rather than having to contract on a programme-by-programme basis. Indeed in a draft of its 1989 SCI TVNZ indicated that it would seek a form of bulk contract. Mounter's view was that programme-by-programme contracting would lead inevitably to disputes over issues of editorial independence and questions of who was to control the programme format, the time the programme was to go to air and the form of the on-screen credits. With the commission interpreting the Act as requiring programme-by-programme funding and insisting on detailed budgets for programmes, disputes were inevitable. Brian Corban remembers that there were many disagreements:

> Over the detail of not just the subject, but also how the programme was going to be produced, who it was to be produced by, what it was going to cost, when it was going to be screened, what size the acknowledgment to New Zealand on Air would be, whether TVNZ should have an

acknowledgment, who owned the copyright in the programme and on and on it went.

However, the Broadcasting Commission had clear obligations under the Broadcasting Act and took a hard-headed and independent approach. It was set up to operate in a fully commercial market and these comments need to be evaluated in this light.

Outcomes of the Separation
One of the primary objectives of setting up the Broadcasting Commission as a specialist purchasing agency was to separate social from commercial objectives and to clarify the responsibility and accountability for each. In this respect at least, the new institutional arrangements appear to have achieved their objective. Both social and commercial objectives were pursued with far greater clarity and transparency than was the case under the BCNZ.

One has only to compare and contrast the annual reports of the BCNZ with the annual reports of the Broadcasting Commission together with those of Television New Zealand Limited to see the difference. Whereas it was impossible to tell from the BCNZ reports exactly how the public broadcasting fee was spent within the organisation, it is now clear from the annual reports of the Broadcasting Commission.[17] In addition to a full set of financial accounts, these reports detail expenditures against each of the five objectives set by the commission (listed above) and provide details of funding by programme type, programme title, broadcaster, and producer. To claim, as do some critics of these reforms, that accountability has actually declined, has little basis in fact.

Following a fee increase from $71.50 to $110 on 1 July 1989,[18] $81.9m was raised from the broadcasting fee in 1990, $83.7m in 1991 and $83.6m in 1992.[19] Of these amounts $31.7, $34.3, and $34.2m in 1990, 1991, and 1992 respectively were spent on assisting television programme production. In 1991 and 1992 approximately 47 percent of the total was spent on the production of Maori, children's and young persons', and special-interest programming. The remaining 53 percent was spent on grants to assist the production of documentary and drama programmes. The hours of programme production the commission has been able to assist rose from 534 hours in 1990 to 944 hours in 1992.

Of the amounts spent by the commission to subsidise television programme production, TVNZ Ltd's direct share of the total amount available fell from 64 percent in 1990 to 35 percent in 1992. However, this understates the benefits which accrue to TVNZ which also hires out its production facilities to those independent producers who are the recipients of New Zealand on Air subsidies. Over this same three-year time period, TVNZ was the designated broadcaster for roughly 77 percent of all television programmes produced with the assistance of commission grants.

Even though the boundaries of the contracting relationship between the commission and TVNZ settled down somewhat after a vigorous start, disagreements caused by the wording of the Broadcasting Act and the commission's interpretation of its role continued. This is reflected in the following personal opinion given to us by Julian Mounter shortly after his resignation as chief executive:

I think, to be honest, that television in this country still under-serves minority and specialist interests and that is a function of the SOE model and how the Commission has come out. It is a consequence of the two together, the combination of the two, that is not working correctly. If the Commission worked totally correctly [in the way I believe it should] the problems would not be there.

What the functions of the commission as a purchasing agency with social objectives should be is, at its most basic level, a political decision. However, it is important that the political debate be properly framed. There is a tendency on the part of some commentators to direct most of their criticism about a perceived lack of minority and special-interest programming (as opposed to popular programming) at TVNZ, claiming that its status as a profit-driven SOE causes it to 'shirk' its responsibilities to cater for narrowly focused audience interests. This and similar criticism seems misplaced. The need is for the debate to be broadened to include an explicit consideration of the role of the Broadcasting Commission and how the public broadcasting fee should be spent. This is a political judgement about social objectives. Although an important source of funding, the commission provided only a small portion of the amounts that TVNZ (and TV3) actually spent on producing local programmes each year.

As a specialist purchasing agency which assists the government in achieving social objectives in broadcasting, the Broadcasting Commission is a unique institution in world broadcasting. Even though there was disagreement about its role and the social objectives it should pursue and the project-by-project nature of its funding, in most respects the formation and operation of the commission served New Zealand taxpayers and audiences well. It opened up the process of decision making and clarified accountability for the spending of public monies to achieve social objectives in television broadcasting.

Political and Social Pressures on TVNZ

As the most powerful and influential of all the media, television broadcasters in New Zealand are subject to almost continuous public and political scrutiny. The resulting pressures are reflected in political tinkering with the structure and operation of state-owned television. As indicated earlier, since the introduction of television in New Zealand in 1960, every national election, with the possible exception of that of 1990, has resulted in the incoming government making changes to the structure of television broadcasting.

The main political and social issues impinging on the operation and management of TVNZ as an SOE have centred on the mix and quality of programmes broadcast, editorial issues, the ownership and orientation of TVNZ's two channels and the progressive removal of legislative barriers to foreign ownership.

Programming Mix and Quality

Since the establishment of TVNZ as an SOE, a number of commentators and politicians have decried the commercialisation of television, arguing that the drive for profits by TVNZ Ltd has debased the mix and quality of programming offered to viewers.[20] In essence, these people argue that, since TVNZ became an SOE, television programming has become 'ratings driven' and that, as a result, there have been significant reductions in the quantity and quality of news and current affairs and in the mix and quality of locally produced programmes which reflect New Zealand's identity and culture.

A thorough assessment of this issue is beyond the scope of this study.[21] It is a matter on which there are many opinions. As Trevor Egerton, who was chief executive of TV3, once pointed out, 'every New Zealander is a programme director with different expectations'. However, the point is that these opinions, even if they are driven by influential elites who believe in giving other people what they think they should want irrespective of whether they want it or not, can translate quickly into political pressure and even party policy. To illustrate, the National Party's policy on broadcasting, released on 16

October 1990, took up the theme that programming on television was becoming 'increasingly sterile and lacks sufficient local social perspective' and that 'many of those programmes which are most important to New Zealand as a society have been sacrificed in favour of those whose only purpose is to maximise the flow of advertising dollars'.

To ensure that state-owned broadcasters act in a socially responsible manner to provide a wide spectrum of public programming, including educational, cultural, entertainment and current affairs programmes, National's policy called for TVNZ to transform the role of Television One into that of a public service broadcaster, emphasising 'quality' programming, and for a National government to give consideration to the sale of Channel 2 to the private sector. Although neither of these policies was actually visited on TVNZ by the National government after it was elected to power in 1990, it does show how vulnerable television broadcasters are to social and political pressures.[22]

As has been alluded to elsewhere in this study, TVNZ's managers argued that they never abandoned the ideals of public service television, even though during the transition period difficult trade-offs had to be made in the interests of creating a commercially viable enterprise. One area which Julian Mounter now admits suffered in the lean years was current affairs programming. However, in the longer term, quality local programming was seen as a critical factor in attracting audiences. On this Mounter commented:

> In the end I believe philosophically that to be an effective broadcaster and commercially viable, you have to have a decent amount of local content. In ten years' time CNN will be producing most of the news, ESPN most of the sport and the entertainment channels most of the entertainment. If you want to be effective as a local broadcaster you are going to have to look local. In the long term it is the local broadcasters' only defence against getting swamped and buried.

A major problem is finding funds for high-cost local production.[23] TVNZ tackled this problem by concentrating on bringing down the cost of local productions and by seeking other sources of funding. The institutional mechanisms for the government to put more funds into local programming to achieve social objectives already exist in the form of the Broadcasting Commission. Although there have been some calls for the imposition of local content quotas on New Zealand television broadcasters, such as those which operate in Australia, Canada, and many European countries, these have not been imposed. There appears to be a shared view at both TVNZ and the Broadcasting Commission that quotas are simply a form of a tax on television broadcasters and that there are better ways of delivering the desired outcomes. Ruth Harley, the executive director of the Broadcasting Commission, made this point as follows:

> A quota is really another name for money. Whose money is the only question. In New Zealand we use a subsidy mechanism; in Australia, it is a levy on the commercial networks. If it is decided that more money has to be spent on New Zealand programmes there are simpler and more effective ways of delivering the result than by quotas. If we are going to say that the commercial networks in New Zealand have to make more New Zealand programmes than they would choose to make on commercial grounds, then what you do is provide them with the funds through contracts for specified services.

Under the current model, the extent of local programme production becomes a political question related to how high the flat tax on owners of television sets (which is what the licence fee essentially is) should be. New Zealand's relatively small population makes funding a fully non-commercial service with substantial local programming very difficult. That is why New Zealand television historically has followed the path of commercial funding through advertising from the beginning.

Beyond the problem of finance there is a second problem of finding sufficient numbers of talented people. Because New Zealand's market is so small, it is difficult to hold on to enough talented writers, actors, and broadcasting journalists, for they tend to move overseas, especially to Australia.

Even with these problems New Zealand content on television rose substantially following deregulation and the establishment of the Broadcasting Commission. Table 7.1 reveals that:

- In total, New Zealand content increased 170 percent from a base of 2,114 hours in 1988, the year prior to deregulation, to 5,714 hours in 1992.
- In prime time (6.00 p.m. – 10.00 p.m.), New Zealand content increased 139 percent from a base of 686 hours in 1988 to 1,640 hours in 1992.
- As a percentage of total hours broadcast, New Zealand content in-

creased 26 percent from 23.9 percent of the schedule in 1988 to 30.2 percent of the schedule in 1992.

These aggregate increases were not confined to one or two genres such as sports and news and current affairs. As can be seen from table 7.1, significant increases took place in all genres with the exception of the information category which recorded only a small increase. New Zealand on Air attributed these increases to growing income from the broadcasting fee and to a much stronger commitment of TVNZ and TV3 to the broadcast of programmes with New Zealand content. [24]

TABLE 7.1: TOTAL LOCAL CONTENT BY HOURS

	1988	1989	1990	1991	1992	% Change 1988–1992
Drama/comedy	39	59	55	86	223	472%
Sports	509	691	1,653	1,283	1,735	241%
News & current affairs	550	709	997	924	1,009	83%
Entertainment	292	458	528	525	886	203%
Children's	325	440	534	739	1,264	289%
Children's drama	12	21	25	20	33	175%
Maori	131	144	143	111	163	24%
Documentaries	43	36	107	139	175	307%
Information	213	253	208	213	226	6%
Total: Prime time	686	943	1,189	1,281	1,640	139%
Total: NZ content	2,114	2,811	4,250	4,040	5,714	170%
% of schedule	23.9%	31.8%	24.2%	31.7%	30.2%	26%

Notes: 1. TV3 commenced in November 1989
2. Figures have been rounded

Source: New Zealand on Air, *Local Content Research New Zealand Television*, Wellington 1988, 1989, 1990, 1991, and 1992

The extent of TVNZ and TV3's commitment to local programming is reflected in figure 7.2 which shows a dramatic rise in the local content screened on TVNZ's two channels and TV3 over the 1989–1992 time period. Panel A shows the increases in total hours of New Zealand content screened by each channel. Panel B shows the signifi-

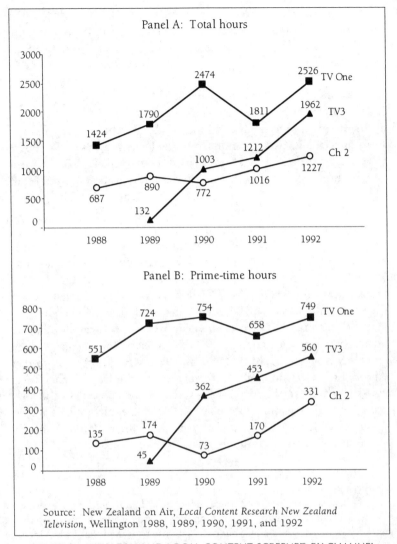

Panel A: Total hours

Panel B: Prime-time hours

Source: New Zealand on Air, *Local Content Research New Zealand Television*, Wellington 1988, 1989, 1990, 1991, and 1992

FIGURE 7.2: NEW ZEALAND LOCAL CONTENT SCREENED BY CHANNEL

cant increase in New Zealand content screened by each channel during prime time.

Well-made local programmes tend to rate highly with New Zealand audiences. It was mentioned earlier that TVNZ's managers argued that a high level of local programming not only rates well but is an important way that TVNZ can differentiate itself from an international competitor who may broadcast into New Zealand in the future. A commitment to a reasonable level of local programming makes sense within the context of TVNZ's commercial obligations and its strategic direction.

Editorial Issues

Independence in programming and editorial matters is important for all media organisations. If anything, visible editorial independence from government is even more important when a country's largest television broadcaster is a state-owned enterprise. This is a critical constitutional issue. The ability of the programme and editorial staff of broadcasting enterprises to make independent editorial judgments is essential to the proper functioning of a democratic society.

Programme and editorial independence has important implications for both organisational and institutional design. First, the chief executive of TVNZ, unlike that of any other SOE except Radio New Zealand Ltd, has a dual role as chief executive and editor-in-chief. Organisationally, this requires a clear separation to be made between the responsibilities of the board and the chief executive on editorial matters. In this respect the chief executive of TVNZ rather than the board controlled the appointment of the most senior managers in the company, and the board did not have direct input into day-to-day editorial matters, although board members were free to make their views known.[25] In the final analysis, the board's power on editorial policy rests on its ability to replace the chief executive if it seriously disagrees with his policies as editor-in-chief. Brian Corban explains his position on editorial independence as follows:

> In the period I was chairman, I don't believe that the news division of the company, or the chief executive as editor-in-chief, would have felt any limitation on their editorial independence and integrity. As a lawyer with an interest in constitutionality I made it absolutely clear that as chairman I would never interfere with the news and that I would fight for their freedom and independence exercised responsibly.

From a wider institutional perspective there was intense political and social interest in the programmatic and editorial judgements made by TVNZ Ltd as television broadcaster and as an SOE. During the establishment of the two media SOEs the issue of their future editorial independence was one of the first questions addressed by the Ministerial Advisory Committee.[26] There were two primary concerns, the first relating to the powers of the shareholding ministers to issue directions to SOEs under section 13 of the State-Owned Enterprises Act 1986 and the second relating to the powers of the shareholding ministers to appoint and remove board members.

In a paper prepared for the MAC in October 1988 Treasury officials pointed out that the powers of shareholding ministers to direct SOEs under section 13 of the Act were severely limited and that any such direction had to be tabled in Parliament within 12 working days. On the issue of board appointments, Treasury officials argued that the ability of shareholding ministers to appoint and dismiss board members was critical to holding directors accountable for their commercial performance, particularly when other means of ensuring performance (such as equity participation by directors) were not available. Officials argued that in the final analysis it was 'the integrity of directors and the nature of their appointment negotiated with the shareholder [that] are important in ensuring editorial independence'.

However, the MAC was not convinced by Treasury officials' opinion that there was 'a large measure of assurance that editorial independence would not be breached' and sought clearer and more explicit protections. In response to these concerns the State-Owned Enterprises Amendment (No. 4) Act 1988 included a section that directly prohibited any direction by the shareholding ministers on programmes, allegations or complaints against programmes, the gathering or presentation of news and/or current affairs programmes, or the responsibility of the company for programme standards. Further safeguards were put into place by inserting appropriate wording regarding independence in programme and editorial matters in the Statement of Corporate Intent, in board members' letters of appointment and in the licence agreement under which the company was forced to operate the broadcasting assets as a result of the Maori Council proceedings.[27]

The only institutional mechanisms in place to deal with programme and editorial standards are those general provisions in the Broadcasting Act 1989 that set out the responsibilities of broadcasters for maintaining programme standards with respect to such matters as good taste and decency, maintenance of law and order, privacy of individuals and the principle of balance when reporting on controversial issues of public importance; the process by which complaints against broadcasters are to be dealt with; and the appointment of an independent Broadcasting Standards Authority to receive and determine complaints against broadcasters.

The importance of clear separation on editorial issues was perhaps best demonstrated in respect of the 29 April 1990 edition of TVNZ's *Frontline* programme. An item in this programme 'For the Public Good' made serious allegations of corrupt links between big business and the Labour government and specifically in terms of the funding of the 1987 election campaign and in the sale of state-owned enterprises. Ministers, Treasury officials and business people named in the programme responded to these allegations with a flurry of complaints and defamation suits (reported to total $7m) as TVNZ came under a barrage of criticism for breaching professional standards of journalism.

Soon thereafter, questions were raised as to whether the Labour government through the shareholding ministers was applying direct political pressure to TVNZ Ltd as 'punishment' for the programme. This was categorically denied in Parliament by the Deputy Prime Minister, Helen Clark, who correctly pointed out that ministerial direction to the company about any programme, allegation, or complaint was specifically excluded by legislation. Subsequent inquiries by the then leader of the opposition, Jim Bolger, raising the question of whether the shareholding ministers had raised their dividend requirement in response to the *Frontline* programme, were found by the Ombudsman to be without foundation.

Instead, the furore over TVNZ's alleged breach of journalistic standards was handled in three ways. First, the chief executive undertook his own internal investigation with the support of the company's chairman, who confined his activities to defending the constitutional responsibility of the chief executive as editor-in-chief to deal with the matter. Based on the chief executive's finding that internal procedures for clearing news and current affairs programmes containing potentially controversial and/or defamatory material had been breached, the executive producer, producer and journalist involved were suspended and the jobs of the producers advertised. The chief executive was quoted in a story in *The Dominion,* 19 May 1990, as admitting that 'the programme fell below the standards required of TVNZ journalists'. Second, the Broadcasting Standards Authority acted on complaints it had received from the Business Roundtable and Treasury. In December 1990, the authority upheld these complaints for factual inaccuracy, lack of balance and deceptive programming practice and imposed a set of sanctions including the screening of apologies and the elimination of all advertising on Television One on the evening of 3 February 1991. Third, individual defamation actions against TVNZ Ltd have continued through the courts.

Although the company weathered the immediate storm of criticism from politicians and others about the *Frontline* programme, this episode did little to improve TVNZ's relationship with the government of the day. TVNZ managers believe some senior Labour politicians were so aggrieved that they took every opportunity to attack the company, the quality of its programming and the integrity of its news staff. The continuing risk to TVNZ Ltd as an SOE is that a high-profile editorial issue involving politicians such as *Frontline*'s 'For the Public Good' could easily have become a focal point for a major swing in attitudes towards the structure and regulation of television broadcasting.

Questions about Privatisation
Suggestions that Channel 2 should be considered for a separate sale to private interests (as recommended by the 1990 National Party policy) were viewed with concern by TVNZ's managers. The reasons were straightforward. The first was that TVNZ's broadcasting strategy centred on its ability to programme its two channels in a complementary fashion. Splitting the two channels apart through sale of Channel 2 would not only stop TVNZ from pursuing this strategy, it would also result in a loss of benefits from shared resources and reduce the attractiveness of TVNZ's remaining channel to advertisers. It may also have two other interrelated effects: one involving programming and one involving the problem of the financing of Television One.

If Channel 2 were to be sold without restrictions being placed on the nature of Television One's operations, TVNZ's managers argue that the end result would be a loss of the distinct choices which are now available on TVNZ's two complementary channels. Instead, all three channels would start to chase the same audience through head-to-head programming as happens in other countries. Channel 2 would move towards becoming a complete channel with the addition of expensive news and current affairs coverage. Inevitably Television One would try then to attract some of the audience away from Channel 2 by adding additional entertainment programming and the creep towards the middle ground would continue.

An analysis of the programme schedules of independent Australian television stations distributed by TVNZ Ltd's Planning Department shows the type of head-to-head programming that results in such markets.[28] This analysis shows the race for the middle ground as networks programme 'like against like'. As one senior manager put it: 'That would drive people mad. They would be very, very annoyed when they saw the end result.' Although the sale of Channel 2 probably would be a godsend for TV3's ratings, TVNZ's managers argued that it would cripple TVNZ Ltd and would have a very significant influence on the nature and mix of programming on all three resulting networks. TVNZ Ltd's *1992 Business Plan* sets out TVNZ's views as follows:

> TVNZ has argued against transferring Channel 2 to the private sector because it would substantially reduce the net worth of the company and returns to shareholders, the standard of service and selection of programming available to consumers would decline and the resulting companies would not have the critical mass necessary to compete (locally or internationally) against global communications companies. (p. 25)

Managers argued that it was this critical mass which allowed the company to cover such major events as the Commonwealth Games, the Whitbread around-the-world yacht races, the Olympic Games and the America's Cup. TVNZ Ltd was listed in 1992 as one of the top 25 publicly owned broadcasters in the world. It also ranked as one of the top 100 broadcasters in the world in terms of size and importance.[29] Splitting the two channels apart runs the risk of marginalising both channels.

There was a related proposition that, if Channel 2 was to be sold, then Television One should be repositioned as a public service channel. This repositioning was part of the National Party's 1990 election policy as noted above. There are two problems with this position. The first is one of finance. Channel 2 runs mainly on overseas programming and contributes the bulk of the net income earned from TVNZ's two-channel network operation. TVNZ's managers argue that loss of this income stream would raise serious problems for the funding of Television One, which would also be under pressure to screen even more expensive local and minority-interest programming than it does at present. With high levels of public service programming, it is possible that Television One would slip to third place in the ratings, and suffer in consequence a significant loss of advertising revenue.

An internal analysis prepared for TVNZ's board in July 1990 estimated that operation of Television One as predominantly a public service channel would capture around $30m in advertising revenue but would cost around $180m to operate—an initial shortfall of around $150m or, if Broadcasting Commission funding of around $15m is included, a shortfall of around $135m. To cover this shortfall the licence fee would have to be raised 130 to 140 percent over its 1992 level.[30]

The second problem is that, even if Television One were funded by alternative means, some believe that the channel would not be an effective communicator. Hugh Rennie makes this point as follows:

> It is not enough to make or show a programme—one has to achieve the substantial attention of the intended audience. In communication terms, specialised publicly funded channels such as the ABC in Australia appear to provide worthwhile programming, but are effectively a failure judged in terms of their effectiveness in reaching a mass audience.

A remaining ownership issue involves Broadcast Communications Ltd. Although it has been part of TVNZ's strategy to sell off shares in its subsidiaries, including BCL, for strategic reasons TVNZ would want to continue to retain at least a 51 percent controlling interest. As noted above, BCL has played a key role in TVNZ's strategy to diversify its activities into electronic communications. It is argued by TVNZ's managers that giving up control of BCL would significantly hinder achievement of this objective.

Foreign Ownership

The Broadcasting Act 1989 passed by the Labour government limited foreign ownership to 15 percent for television broadcasters. The National government elected in 1990 soon raised this level to 49.9 percent ostensibly to allow TV3 to restructure and refinance by allowing an expanded foreign ownership. Both TVNZ Ltd and TV3 lobbied hard to protect their interests. In opposition to increased foreign ownership, TVNZ raised the spectre of possible editorial interference by international media interests in New Zealand politics and elections, while TV3 dismissed such fears as groundless. In fact, TVNZ's real concern was competitive in nature. The company was lobbying to protect itself from TV3 being taken over by a multinational media company with access to large programme and film libraries. Were this to happen, TVNZ's managers believed that the financial success and survival of the company would once again be at risk.

Not long thereafter, the Minister of Broadcasting, Maurice Williamson, after publicly arguing against 100 percent foreign ownership, suddenly reversed his position and announced that the government would remove all restrictions on the level of foreign ownership. Again TVNZ found this extremely threatening to their competitive position and protested vigorously. Their protests fell on deaf ears and the Broadcasting Act 1989 was amended in 1991 to allow 100 percent foreign ownership.

The removal of all foreign ownership restrictions fundamentally altered the dynamics of the New Zealand industry and was a matter of grave concern to TVNZ. The company's managers believed they now faced the daunting prospect of a major foreign media owner such as Kerry Packer gaining control of TV3. TVNZ feared that Packer could then use his ownership of Australia's Channel Nine network to buy the Australasian rights for TVNZ's highest-rating programmes and sporting events. TVNZ Ltd under these circumstances would simply be out-gunned. In addition, it would lose valuable contracts with Channel Nine involving the joint use of overseas facilities to TV3. Julian Mounter immediately proposed a strategy to respond to a possible move by Packer into New Zealand television.[31] One of the reasons Mounter had proposed that TVNZ Ltd acquire a share of Channel Nine was to help forestall such a move, a strategy that he believed 'would have prevented what [was now] probably going to happen'.

As it turned out this 'worst-case' scenario did not eventuate. Instead, a stake in TV3 was acquired by CanWest Communications, a Canadian television company. CanWest was not the type of international company that TVNZ most feared. However, this rapid shift in government policy did have an immediate impact on the extent of TVNZ Ltd's holdings in Sky Network Television. As noted in chapter 5, the policy change resulted in Sky's American investors immediately requiring a 51 percent shareholding as a condition of their investment. Julian Mounter noted that the new American partners were originally set to come in with less than a 50 percent holding but

. . . right at the wrong moment [for TVNZ] the government allowed 100 per cent foreign ownership and the American companies just turned around and said 'We are not coming in unless we get 51%'. We were over a barrel.

Mounter believes that the later altercation with Sky over rights to the South African rugby tour by the All Blacks could have been avoided if the New Zealand shareholders had retained control. The problem in his view was that the American shareholders who had control did not understand that 'rugby is a religion in New Zealand'.

Conclusion

When TVNZ became an SOE, a number of internal and external relationships became increasingly critical and contentious. Within TVNZ the governance relationship between the board and management, in particular between the chairman of the board and the chief executive, was especially significant. The first chairman of TVNZ Ltd, Brian Corban and the first chief executive, Julian Mounter, established a strong, positive working relationship which facilitated TVNZ's successful transition.

There were difficulties, however, in defining the board's role given the unusual terms of appointment of the members of the first board. This complicated the governance process in the first year of TVNZ's life as an SOE with some board members finding it difficult to reconcile the requirements of the State-Owned Enterprises Act with their letters of appointment. Time and energy that could have been spent on overseeing the company's preparation for competition was instead invested in board discussion and in increasingly heated disputes with shareholding ministers over the primacy of commercial or social objectives as set out in the company's Statement of Corporate Intent. Subsequent board appointments were made to strengthen the commercial orientation of the board and to improve its ability to govern the company.

At least during the transition phase, TVNZ's relationship with its shareholding ministers and their advisers was stormy and contentious. In part this was due to difficulties created by the wording of the first board's letters of appointment, but it was aggravated by poor financial reporting systems within TVNZ and ongoing disagreement over what the company's rate of return and dividend payouts should be. To some extent, these tensions also reflected the fact that the company and the government as owner were driven by different imperatives. The more commercially oriented members of TVNZ's board and management believed that it was necessary to protect their markets and take risks to secure a viable financial future for the TVNZ group. On the other hand, shareholding ministers and their advisers wanted to ensure that the board limited the government's exposure to risk. Given the choice between the use of their investment to generate prospective (and risky) returns or the certainty implied by the payment of dividends, shareholding ministers were consistent in expressing a strong preference for profits and dividends sooner rather than later.

Although these differences resulted in strains in the relationship, they served the purpose of improving the shareholders' understanding of the business and helped to focus TVNZ's board and management on the need to meet rate-of-return and dividend expectations.

The other external relationship of particular consequence was with the Broadcasting Commission, or New Zealand on Air, as it soon became known. As a new organisation charged with promoting the government's social objectives in broadcasting as defined in the Broadcasting Act 1989 and funded by the public broadcasting fee, tension between the commission and TVNZ was inevitable, at least during the settling-in period as both parties attempted to define the new relationship. In the old BCNZ days, all the broadcasting fee had gone to the public broadcaster; now it was to be divided up among public and private broadcasters and private production houses. Thus, the former monopoly had to come to terms with a new competitive environment for public funding as well as for advertising dollars. Furthermore, Julian Mounter and TVNZ had certain expectations as to how the commission would operate and what types of programmes it would fund. These expectations did not always accord with what Ruth Harley and the commission had in mind. These differences were highlighted when the commission funded several TV3 programmes that Mounter viewed as commercial in nature. But such funding reflected the commission's own interpretations of its legislative charter and its strategy of encouraging and developing New Zealand programming in general, rather than what might be of interest to small elite audiences.

Both TVNZ and the Broadcasting Commission were seeking to develop new strategic directions and to carve out new roles under strong leaders. Not only was some conflict inevitable but it also had a useful purpose as it helped each party to clarify its focus and direction in what was a new and emergent industry environment.

In addition to tensions that resulted from these external relationships, TVNZ was also subject to a wide variety of political and social pressures after becoming an SOE. These pressures had also affected the old BCNZ as television broadcasting had always been the subject of social and political debate. But the pressures were exacerbated by corporatisation and SOE legislation which required TVNZ to explicitly focus on commercial objectives and profitable performance. Debates swirled around TVNZ Ltd from the outset over programme mix and quality, editorial matters, foreign ownership issues, the future ownership structure of TVNZ's two channels, and whether New Zealand should have a 'public service' station. Over the period of our study, these pressures were constant and unrelenting and demanded considerable board and management time. While coping with these types of pressure is a fact of life for any media company, it was particularly so for TVNZ because it had come to be so much in the public eye as a result of its establishment as an SOE.

CHAPTER EIGHT

Financial Performance

Establishment as a state-owned enterprise placed new importance on the financial performance of TVNZ. It had to operate in a more transparent environment, with clearly specified and public financial objectives spelled out in its Statement of Corporate Intent. Shareholding ministers were adamant that TVNZ's financial performance must be as good as those of comparable businesses not owned by the Crown and that a satisfactory dividend must be paid. And TVNZ management was very conscious of the potential impact on profitability of the entry of a powerful competitor in the industry. The financial performance, then, of TVNZ since corporatisation is an important indicator of the company's successful transformation into a commercial entity.

In this chapter, we examine the impact of competition on TVNZ's revenue streams and look at its changing revenue mix subsequent to the establishment of the Broadcasting Commission and changes in the allocation of the public broadcasting fee. We also analyse TVNZ's profitability over the four years of our study and the level of its dividend payments to the shareholder over this time period. And since the New Zealand state-owned enterprise model requires SOEs to be as financially successful as comparable companies in the private sector, we compare TVNZ's performance to that of other companies both within the broadcasting industry and in other industries.

Analysis of Performance

This analysis is based on the financial results of the BCNZ for the 1979–1988 period and of TVNZ Ltd for the 1989–1992 period. Financial data are presented in the appendix and include summary income statement data, balance sheet data, and a comprehensive set of financial ratios, the major ones of which are analysed below.

There are a number of issues that affect the degree of confidence it is possible to place in the analysis and any conclusions drawn are inevitably tentative. First, BCNZ had a monopoly on television advertising revenues, faced no competition for programme acquisition and received the entire public broadcasting fee (PBF), whereas TVNZ Ltd has faced competition in each of these areas. Second, the results for the BCNZ include the activities of Radio New Zealand, the *New Zealand Listener* and the Symphony Orchestra as well as Television New Zealand. Unfortunately, there were no data available that enabled us to separate out performance measures of TVNZ as a division of the BCNZ apart from some pro-forma data that were published as an appendix to the Rennie Report.[1] The situation was nicely summarised in the Treasury report on the valuation of TVNZ as follows:

> The BCNZ has had a history of operating as a commercial entity and a provider of social and/or non commercial services. In addition TVNZ has operated with no direct competitor and RNZ has enjoyed a dominant though diminishing market position. Due to the corporate structure the allocation of overheads for the provision of transmission and other corporate services, such as accounting and treasury functions, has been arbitrary. In addition the accounting treatment of income received from the public broadcasting fee has not been consistent in the past decade. Accordingly, it is not possible to rely on the service's profit history as an indicator of future performance.

Third, in the TVNZ Ltd period there were two major changes in accounting methods that affect the analysis. Firstly, the corporation changed its method of accounting for locally produced programming that was in inventory at year end. The impact in 1990, the year the change was made, was to increase after-tax reported income by $2.28m. Secondly, TVNZ ceased to equity account the *New Zealand Listener* and Sky in 1990 and CLEAR in 1991.[2] No attempt has been made to adjust the financial reports for these changes as the alternative numbers were generally not available. In any event almost all the analysis in this chapter is undertaken on the income numbers excluding the profits and losses of associate companies, so the only item that would have an impact on the analysis is the change in valuing inventory in 1990. TVNZ Ltd also reported extraordinary gains of approximately $15m during the 1989–1992 period, which have been excluded from the formal analysis that follows.[3]

Changes in Revenues and Expenses

Figure 8.1 shows total television advertising revenue over the 1979–1992 period. Advertising revenue increased rapidly over the period as television captured a larger share of the advertising market, mainly at the expense of newspapers. The 1989 period is for thirteen months[4] while the results for 1990 show the effects of the first full year of competition from TV3 on TVNZ's advertising revenues.

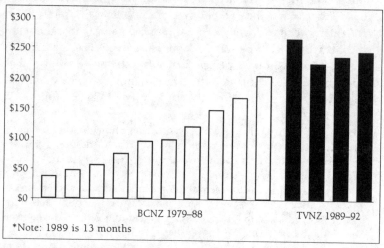

FIGURE 8.1: TOTAL ADVERTISING REVENUE ($M)

Average advertising revenue per employee rose from $64,200 for the BCNZ era 1984–1988 to $129,100 for TVNZ over the 1989–1992 period. This increase indicates a substantially higher level of productivity per employee.[5] In the TVNZ period, the revenue trend has been upwards except for 1990, the first full year of competition. A graph showing the increase in advertising revenue per TVNZ equivalent employee over the 1984–1992 period is shown in figure 8.2.

TVNZ's managers estimated that they would lose around $30m of annual advertising revenue to TV3. In its prospectus issued in November 1989 TV3 predicted that it would capture in excess of $100m of advertising revenue in their first full year of operations. Part of this $100m would come from advertisers switching away from TVNZ to TV3 and part would be 'new' revenue coming from advertisers who were not then using television because television advertising time was already fully sold at peak times.

TVNZ's estimate of loss of advertising revenue to TV3 was much

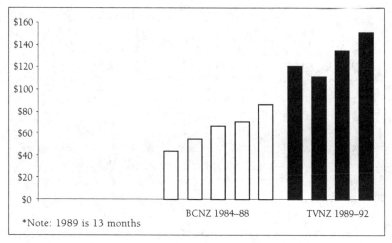

FIGURE 8.2: TELEVISION ADVERTISING REVENUE
PER TVNZ EQUIVALENT EMPLOYEE ($000)

lower and much closer to the correct impact of TV3's entry. By transforming its sales and marketing activities, TVNZ managed to maintain its advertising revenues around the $250m mark during a period in which the general economy was in recession and the advertising market as a whole was flat.

The small size of the television advertising market in New Zealand and the entry of TV3 raise the question of whether there was enough room for three profitable free-to-air channels in New Zealand. To some extent this room has been created by a growth in overall spending on television advertising and a decrease in TVNZ's advertising revenues. However, in the lean years of 1990 and 1991, when the general economy was in recession, the competition for advertising revenue became a matter of survival, as was demonstrated by TV3's inability to achieve the advertising revenues it predicted and subsequently going into receivership.

The restructuring of the broadcasting industry and TVNZ's strategy of seeking other sources of revenue is demonstrated in the company's changing revenue mix over the fourteen year period examined here. In 1979 the licence fee accounted for approximately 55 percent of combined television and radio advertising revenues. By 1992 the amount of the public broadcasting fee (PBF) received by TVNZ from the Broadcasting Commission was about 8 percent of the company's advertising revenues. Although it remained a material and important source of revenue, the amounts received from the PBF became far less critical to the financial well-being of TVNZ. Since TVNZ has been corporatised Other Income (Revenue) has risen from $30m (11.5 percent of advertising revenue) in 1989 to $76m (31 percent of advertising revenue) in 1992. The other income consisted of revenue from transmission services, outside broadcasts for third parties and the sale of television programmes.

In future years the TVNZ group expected to add income from its investments in Sky and CLEAR. In this respect, TVNZ Ltd's strategy was to hold on to its share of audiences and advertising revenues against competition, at least until the benefits of its investments in these areas begin to flow. In the company's view, maintaining its two-channel share of audiences in the 75–80 percent range depended on it retaining important prime-time programmes, staying ahead of its competition with its promotions and marketing, improving the quality and ratings of its local programme production, and, most importantly, retaining critical mass through the continued control of its two channels.

One interesting feature of the BCNZ's accounts that was previously mentioned in chapter 7 is that it was not until 1987 that the BCNZ disclosed any information in its annual report about how it allocated the PBF within the corporation. In a footnote to the 1987 accounts the corporation disclosed the amounts allocated to the Symphony Orchestra, National Radio, and the Concert Programme, remote area transmission and television programme production. At that stage television programming and production accounted for approximately 18.3 percent of the total net of collection costs. In 1988 that percentage rose to 27.7 percent. The annual accounts of the Broadcasting Commission since its establishment show that the average proportion of the PBF spent on television has been about 45 percent of the total fee net of collection and administration costs.

Total expenses increased on average at a rate of 15.5 percent per annum for BCNZ during the 1984–1988 period and at a rate of 3 percent per annum once TVNZ was corporatised.

Changes in Profitability and Dividend Payouts
Of more interest are TVNZ's profitability figures which are shown in figures 8.3, 8.4, and 8.5. Figure 8.3 graphs Earnings Before Interest and Tax (EBIT), figure 8.4 graphs Return on Assets (RoA), and figure 8.5 shows Return on Equity (RoE).

FINANCIAL PERFORMANCE

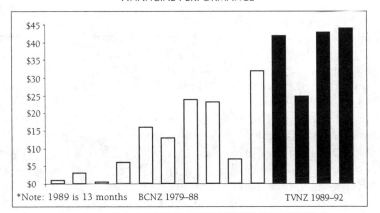

FIGURE 8.3: EARNINGS BEFORE INTEREST AND TAX ($M)

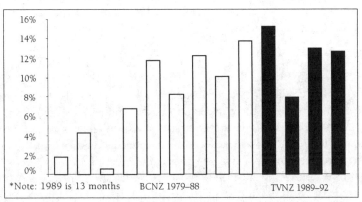

FIGURE 8.4: RETURN ON ASSETS (%)

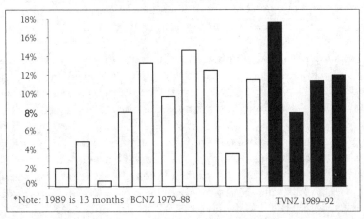

FIGURE 8.5: RETURN ON EQUITY (%)

While all three graphs show 1990 results to be well below those for 1989, 1991, or 1992, the general conclusion is that the level of profitability of TVNZ Ltd post-corporatisation was substantially above that achieved by the BCNZ. To illustrate, RoA was an average of 8 percent for BCNZ and about 12 percent for TVNZ Ltd. As it turns out, the equivalent numbers for RoE were also 8 percent and 12 percent respectively. However, relative to TVNZ Ltd, the BCNZ numbers for RoE are both inflated and deflated. First, the numbers for the BCNZ era are inflated relative to the equivalent TVNZ Ltd because no taxation expense is included in the BCNZ results. Second, the numbers for the BCNZ era can be considered to be deflated relative to the TVNZ Ltd numbers because the BCNZ was financed almost entirely by higher-cost equity. TVNZ Ltd has financed its assets with an approximately equal mix of debt and equity and the return on assets has been greater than the interest cost.

The ratio of Earnings Before Interest, Depreciation and Taxation to Total Income on average was 10.6 percent in the BCNZ era and 16.1 percent for the first four years of the life of TVNZ Ltd. The corresponding EBIT/Income ratios are 5.5 percent and 12.1 percent respectively. Part of the explanation of this increase reflects the fact that the BCNZ had television and radio divisions. Television is a more capital-intensive activity than radio and hence would be expected to show higher margins when the averaging effect of radio is removed from the figures. To illustrate, the Sales/Assets ratio of BCNZ was 128 percent. For TVNZ Ltd it has averaged 98 percent. However, the Sales/Asset ratio increased over the period 1990–1992 at a time when margins initially increased and then remained constant. A plausible explanation for the change in this ratio is that TVNZ Ltd was working its assets harder.

Over 1989–1992 TVNZ Ltd provided and paid dividends of $37.75m to its shareholding ministers. Over the preceding ten-year period 1979–1988 the BCNZ paid dividends of $7.268m. A provision of $15.5m for dividends for 1988 is shown in the accounts of the BCNZ at 31 March 1988. This was the amount the Minister of Finance directed the BCNZ to pay under the Broadcasting Act 1976. Only $6.268m was paid as the corporation's board disputed both the legality and commercial practicality of the level of the proposed dividend. The remaining $9.232m was never paid as the dispute was made moot by the establishment of TVNZ Ltd as an SOE.

In summary, subject to the caveats expressed above and an additional one that will be mentioned below, it is reasonable to conclude that TVNZ in the period 1989–1992 reached a level of profitability about 50 percent higher than did the BCNZ even though it had to operate in a competitive environment. The additional caveat is that it was highly likely that TVNZ, as a division of the BCNZ, was cross-subsidising the radio division. Nevertheless, we would argue that there is strong evidence of an improvement in financial performance from the BCNZ era to the TVNZ Ltd era. There is also quite strong evidence that TVNZ Ltd has improved its financial performance over the period 1989–1992.

Changes in Financial Structure
One other important change pre- to post-corporatisation involves leverage. The BCNZ was financed almost entirely by equity, carrying very little debt throughout much of its life even though the Broadcasting Act of 1976 allowed the corporation to borrow. Until the final year of the BCNZ's existence, when the Auckland Television Centre was built, shareholders' funds never fell below 80 percent of total assets. In the first year of its existence TVNZ Ltd put a long-term borrowing facility in place, and shareholders' funds dropped to about 55 percent of assets. As noted above, this change in leverage contributes to the increase in RoE from the BCNZ era. One issue is whether this increase is sufficient to offset the additional financial risk associated with a higher level of debt in the capital structure. It is reasonable to conclude that the increase in RoE has been sufficient given that the new level of leverage would not be regarded as excessive by most standards. Leverage was substantially less than that reported in *Value Line* for the Broadcast/Cable TV industry in the United States over the 1988–1992 period (an average long-term debt to common equity ratio of 251 percent). TVNZ has carried a larger debt than was originally forecast because it was unable to dispose of surplus assets or sell shares in its subsidiaries due to constraints in TVNZ's licence agreement.

The final aspect of balance sheet structure that deserves attention is the ratio of current assets to current liabilities. In the BCNZ era this ratio averaged 2.5:1 and was boosted in the mid 1980s by large amounts of cash being carried by the corporation. In TVNZ the ratio from 1989–1992 was about 1.8:1 with no cash balance but substantially higher inventory levels (made up primarily of programme inventories). Between 1988 and 1990 inventory levels doubled, involving significant cash outflows. The large cash outflow associated with this major build-up in programme inventory was not fully captured in the projected cash outflows used for the initial discounted cash flow valuation undertaken in 1988.

Other Performance Comparisons

Performance Compared to Targets

Another way to evaluate TVNZ Ltd's financial performance is to compare rates of profitability with those targeted in successive Statements of Corporate Intent. The profitability measures used for this purpose are Earnings After Tax to Shareholders' Funds (EAT/SHF, i.e. RoE) and Earnings Before Interest and Tax to Total Assets (EBIT/TA, i.e. RoA).

The results for the one year ahead projections are summarised below in table 8.1. As can be seen by a review of this table TVNZ Ltd has generally undershot its projections but not by much. However, as shown by a review of the summary income statement data in the appendix, actual profits are significantly affected by the inclusion of large extraordinary items in 1990 and 1991.

TABLE 8.1: PROJECTED AND ACTUAL PERFORMANCE 1989–1992

Year	1989	1990	1991	1992
EAT/SHF (RoE)				
Projected	16.2%	9.8%	13.1%	11.6%
Actual	17.6%*	7.9%	11.3%	11.9%
EBIT/TA (RoA)				
Projected	16.1%	9.9%	17.3%	13.6%
Actual	14.9%*	7.9%	13.9%	12.6%
* for thirteen months				

Table 8.2 provides a more detailed comparison of TVNZ's Ltd projected targets contained in their 1990–1992 SCIs compared against actual performance. Study of this table shows a 'hockey-stick' phenomenon in TVNZ Ltd's projections.[6] This phenomenon is seen most clearly by comparing the three SCI projections for 1990, 1991 and 1992. In each case we see that the projections for the second and third year of the three-year prediction period were materially higher than for the first year.

TABLE 8.2: COMPARISON OF SCI PROJECTED TARGETS TO ACTUAL RESULTS

Year of SCI	1989	1990	1991	1992	1993	1994
1989						
EAT/SHF						
Projected	16.2%	20.7%	12.8%	11.7%	12.8%	
Actual	17.6%	7.9%	11.3%	11.9%		
EBIT/TA						
Projected	16.1%	16.2%	12.4%	12.9%	15.3%	
Actual	14.9%	7.9%	13.0%	12.6%		
1990						
EAT/SHF						
Projected		9.8%	13.9%	16.5%	18.3%	
Actual		7.9%	11.3%	11.9%		
EBIT/TA						
Projected		9.9%	19.7%	21.6%	22.1%	
Actual		7.9%	13.0%	12.6%		
1991						
EAT/SHF						
Projected			13.1%	15.2%	18.4%	
Actual			11.3%	11.9%		
EBIT/TA						
Projected			17.3%	19.2%	21.7%	
Actual			13.0%	12.6%		
1992						
EAT/SHF						
Projected				11.6%	13.8%	14.3%
Actual				11.9%		
EBIT/TA						
Projected				13.6%	14.6%	14.6%
Actual				12.6%		

What also is interesting is that the set of projections in each SCI is generally lower than the set made in the preceding year's SCI. One interpretation is that it was a natural response to the considerable pressure from shareholding ministers and their Treasury advisers for improvements in reported rates of return. Given the discussion in chapter 7 on this issue this is certainly a plausible interpretation in the case of TVNZ Ltd. In his final report to the board in September 1991, Julian Mounter wrote:

> Our main failure in this regard was to over-promise in each of the last two Statements of Corporate Intent. In a meeting with Ministers, earlier

this month, [the chairman] and I explained that we believed we should not have allowed ourselves to be 'bullied' into promising such high returns.

The trend in TVNZ Ltd's projections was to progressively lower shareholding ministers' expectations as to what TVNZ's board and managers believed were reasonably achievable targets.

Performance Compared to the Industry
TVNZ's performance can be compared with that of other television operators. First, using data from the United States we can compare TVNZ's financial performance with that of major American networks. Financial data taken from *Value Line* for the 1988–1992 period shows an RoE for two of the major networks (CBS and ABC) of an average of 9.7 percent. TVNZ's average for the same period was 11.8 percent.

We can also compare TVNZ's RoE with the performance of Australian television broadcasters over the 1988–1992 period. From *Jobson's Year Book of Australian Companies* and *Australia's Top 300* we collected data for seven companies involved in television broadcasting, providing us with 24 data points on RoE. Of these 24 data points, nine (37.5 percent) involved losses. The average RoE for the profitable company years was 7.7 percent.

Table 8.3 reports the sequence of aggregated profits (losses) before extraordinary items of commercial television stations in Australia for the period 1987 to 1991.

If extraordinary losses are included in 1990–91, the loss for that year rises to $2 billion. TVNZ earned profits over this entire period.

TABLE 8.3: AGGREGATE PROFITS (LOSSES) OF FIFTY AUSTRALIAN
COMMERCIAL TELEVISION STATIONS IN AUSTRALIAN DOLLARS

1987–88	1988–89	1989–90	1990–91
$141.3m	($3.3m)	($80.9m)	($225.3m)
Source: Australian Broadcasting Tribunal, *Broadcasting in Australia*, 4th ed., September 1992.			

Lastly, the three major Australian commercial television networks were placed in receivership or were subject to a capital reconstruction during the 1989–90 period. The Channel 9 network was the only television network to improve its profits before interest and tax in the major markets of Sydney, Melbourne, Brisbane, and Adelaide. This time period was clearly a difficult one for commercial broadcasters in the region because of the depressed advertising market.

From the limited evidence presented here, TVNZ's financial performance over this period appears stronger than that of similar commercial networks in either Australia or the United States of America.

Performance Compared to New Zealand Listed Public Companies
We also collected summary profitability data on 82 listed public companies that survived through the period 1988–1992. Table 8.4 summarises the results by giving the median and upper and lower quartiles for RoE measures for the companies, with extraordinary items excluded. The table also gives mean RoE numbers after removing some extreme outliers from the data.[7]

TABLE 8.4: SUMMARY ROE MEASURES FOR LISTED
PUBLIC COMPANIES 1988–1992

Year	Lower quartile	Median	Upper quartile	Mean
1988	-3.90%	6.54%	12.13%	5.99%
1989	-3.29%	7.03%	14.41%	7.04%
1990	-6.98%	5.01%	13.49%	.35%
1991	-11.05%	4.96%	11.26%	1.60%
1992	-7.00%	9.11%	13.51%	6.08%

Table 8.5 shows where the RoEs of TVNZ lie relative to the general benchmarks in the preceding table.

TABLE 8.5: COMPARISON OF ROE FOR TVNZ WITH
BENCHMARK ROES FOR LISTED COMPANIES

Year	Lower quartile	Between median & lower quartile	Between median & upper quartile	Above upper quartile
1989				TVNZ
1990			TVNZ	
1991				TVNZ
1992			TVNZ	

In each case TVNZ results lie above the median for all listed public companies. This benchmark is biased upwards as it only includes surviving companies. We appreciate that there has been no attempt made here to factor risk into this analysis although the leverage discussion involves some consideration of such issues. Nevertheless our overall conclusion is that the performance of TVNZ as measured by RoE has been quite strong since being established as an SOE.

Conclusion

Data limitations make it difficult to determine conclusively whether TVNZ was more efficient and profitable as an SOE over the 1989–1992 period than it had been as a division of the old BCNZ. While the data is consistent with a considerable improvement in performance, the quality of the data prior to corporatisation is so poor that any conclusion is necessarily tentative. Despite some slowness to rationalise its staffing structure and to cut costs, TVNZ achieved financial results in the 1989–1992 period well ahead of the performance assumed in its valuation although below the targets agreed in successive Statements of Corporate Intent.

However, from the preceding review of TVNZ's financial performance we can reasonably conclude that the TVNZ group of companies for the period under review has operated as a successful business which is as profitable as comparable businesses not owned by the Crown. This is particularly noteworthy given the new competitive challenge TVNZ faced with the entry of TV3 into the broadcasting market and the poor state of the economy in general and of broadcasting in particular, during the 1989–1992 period.

CHAPTER NINE

Managing Radical Change in
New Zealand Television

The last decade has witnessed remarkable change in New Zealand broadcasting. Following a succession of reports and commissions suggesting future directions for New Zealand broadcasting, TVNZ was established as an SOE with a strong commercial focus. TV3 entered the free-to-air market as a direct competitor to TVNZ and pay television in the form of Sky Television competed for audience share. Regional stations such as Canterbury Television emerged to fill the gaps left by TVNZ's downsizing. New institutional arrangements for monitoring broadcasting standards and for the purchase of New Zealand content were put in place. Technological advances meant that the television industry in New Zealand could no longer be regarded in isolation but rather had to be seen as part of a global electronic communications market.

TVNZ faced a quite different world as an SOE in a deregulated, competitive environment than it had as a sheltered division within the BCNZ. In this chapter we draw some conclusions from the re-making of TVNZ to meet these challenges. Television broadcasting in New Zealand is particularly interesting because, as is shown in figure 9.1, it has been shaped by political and social pressures, rapid technological change and globalisation, the introduction of competition in broadcasting and telecommunications and changes in ownership form and institutional structure. An analysis of TVNZ's experience over the period of its transition into an SOE provides a number of insights about the importance of transformational leadership and the management of radical organisational change. It also provide insights into the process and problems of turning a monopolistic state broadcaster into a successful electronic communications business operating in a competitive environment.

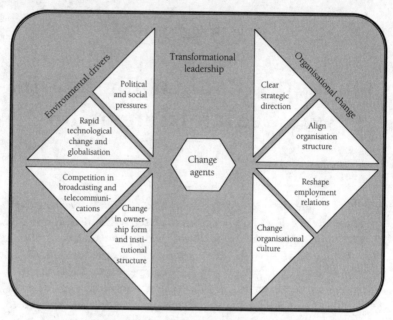

FIGURE 9.1: REMAKING TELEVISION NEW ZEALAND

Environmental Drivers of the Remaking of TVNZ

What stands out in the transformation of TVNZ into an SOE are the profound effects that changes in technology, competition, and ownership structure have had on the broadcasting industry and the management of TVNZ. In a world in which direct broadcast delivery from high-powered satellites became feasible and digital technology was creating a revolution in electronic communications, it was increasingly difficult for the government to continue to try to protect the state-owned broadcaster's monopoly through the political process. Had not the Broadcasting Tribunal issued a warrant to a privately owned third television network, it would have been only a question of time and economics before competing broadcasters using alternative methods of delivery arrived in New Zealand.

By themselves, technological change and the prospect of competition were sufficient to force TVNZ, as a division of the BCNZ, to start the process of change and cause its managers to look outward and prepare for the future. However, it was still hobbled by being part of a statutory corporation operating under an Act of Parliament which resulted in mixed and conflicting objectives, inflexible and bureaucratic personnel systems, and inadequate financial management and information systems. Quite simply, as a statutory corporation the BCNZ lacked a structure in which radical change could occur.

It was the decision to establish TVNZ as a separate SOE which finally gave the managers of TVNZ the ability to respond quickly and flexibly to economic and technological pressures and to properly position the organisation to meet the threat posed by new entries. Had TVNZ remained a division of BCNZ, as part of a statutory corporation operating under the constraints imposed by the Broadcasting Act of 1976, it might not have survived the onset of competition—not, at any rate, in anything remotely resembling its current form.

Competition and rapid changes in technology provided vital catalysts to the expansion of viewing choices available to audiences in New Zealand. The entry of TV3 changed the nature and mix of programming screened by TVNZ, as it now had to compete to hold audiences. Even though TVNZ finally withdrew from regional television (against opposition from politicians and some of its own board members), this simply left a niche in the market for others to fill. With the cost of basic television broadcasting equipment falling rapidly, other broadcasters started to move into the regional market. Together with the introduction of pay television in the form of Sky Television, the choices available to New Zealanders expanded dramatically after legislative and regulatory barriers to entry were removed.

Furthermore, the leaders of TVNZ had to operate within a new institutional framework. At the same time as TVNZ was established as an SOE, a new purchasing and monitoring regime in the form of the Broadcasting Commission and the Broadcasting Standards Authority was formed. Instead of being bulk-funded from the public broadcasting fee, TVNZ faced competition for public money and was funded on a programme-by-programme basis. This resulted in an interesting new balance of power. Although TVNZ remained a dominant player, it could no longer call the tune. As Roger Horrocks, who had been a member of the Broadcasting Commission since its establishment, commented:

> The structure is a very interesting one in terms of New Zealand on Air and the broadcasters. Nobody ever feels they have any power. Everyone feels that in a sense they are constrained by the environment. TVNZ never feels that it has enormous power. They feel that they are satisfying their customers. They feel they are competing with TV3 and dealing with the foibles of New Zealand on Air. New Zealand on Air never feels it has any power. Nothing we want can get on the television unless the broadcasters want it and our money is very limited. Programme makers don't feel that they have any power. That is the interesting thing. It is like a jigsaw puzzle with the pieces interlocking.

TVNZ had to find its feet within this new institutional structure, which placed additional external constraints on their operations. The character of the external environment within which TVNZ operated had changed within the space of a few short years from being very stable and predictable, largely subject to the control of BCNZ, to being unpredictable and very dynamic, with TVNZ as only one player among a number of others. To retain its pre-eminent position in this new environment, TVNZ required radical changes in strategy, structure, and culture.

The Importance of Transformational Leadership

Successful organisational turnarounds need strong transformational leaders who can challenge and lead the organisation through a radical change process. As Tichy and Ulrich said of the role of the transformational leader:

> What is required of this kind of leader is an ability to help the organization develop a vision of what it can be, to mobilise the organization to accept and work toward achieving the new vision and to institutionalise the changes that must last over time.[1]

In the case of TVNZ several different people played a critical role at different stages in shifting TVNZ from its historical track into a new direction. Initially, the challenge to the old ways came with the appointment of Hugh Rennie and a new BCNZ board in the mid 1980s. The challenge to the old organisation intensified in 1986 with the appointment of Nigel Dick as chief executive of the BCNZ and Julian Mounter as director-general. Mounter, in particular, set about persuading people that change was inevitable because it was driven by technological advances and global competition rather than legislation. Brian Corban, as deputy chairman of the BCNZ, then chairman of the Ministerial Advisory Committee and finally as chairman of TVNZ Ltd's board of directors, helped to drive the process of change forward.

Subsequently appointed as TVNZ Ltd's first chief executive, Mounter was the primary architect of change at TVNZ. Drawing on his international experience in the broadcasting industry, he got the process of change started by reorganising the top management team and seeding the organisation with managers with commercial broadcasting experience gained either in the highly competitive New Zealand radio market or overseas. Without a chief executive of Julian Mounter's experience and single-minded drive to ready the company to meet competition from a third network, the outcome might have been entirely different. Mounter communicated a vision of TVNZ's future in global terms and then motivated and drove managers and staff to break through barriers to achieve that vision.

The chairman of TVNZ Ltd also played an important role. As chairman Brian Corban supported and championed his chief executive's strategic vision and worked to ensure that management had the resources and flexibility they believed they needed to build a viable and financially successful electronic communications company. This close working relationship between Brian Corban and Julian Mounter was an important factor throughout the transformation process.

In addition to strong leadership at the top of the organisation, change agents are required to translate the vision of the transformational leader into organisational structure and culture. These change agents are critical for modelling the required changes, assisting in the developments of new structures that will institutionalise the new mission of the organisation and converting others to the changed direction of the company.[2] In the transformation of TVNZ, Chris Gedye, the leased executive who introduced the SBU structure into TVNZ, was a critical change agent, as was Stewart McKenzie in the Finance Department, to whom fell much of the work of making the SBU system work and ensuring that managers and producers paid attention to the bottom line. Don Reynolds, formerly an independent producer, pushed the limits of internal charging as general manager of South Pacific Pictures. The new manager of marketing, Michael Dunlop, introduced innovations from the competitive radio industry that transformed TVNZ's marketing department into a highly professional and competitive force in the industry. John McCready, as director of programming, greatly influenced the on-screen look of TVNZ and its programme schedules. The remaking of TVNZ was assisted greatly by change agents such as these key managers who were brought in from outside the organisation.

There is no doubt that TVNZ's approach to managing the organisational turnaround that its leadership perceived as necessary following corporatisation was revolutionary not incremental, top-down in its direction and directive rather than consultative. Stace and Dunphy suggest that such 'frame-breaking' change is necessary when an organisation is out of strategic alignment with its environment, has only a short time within which to turn itself around and is faced with a lack of support for radical change from within the organisation.[3] These were the conditions facing TVNZ's leadership in 1988 as they approached corporatisation.

However, drawing on their research on organisational change in Australia and New Zealand over the past decade, Stace and Dunphy further suggest that organisations can only survive in this turnaround mode for 2–3 years at most before cynicism and alienation settles in among staff, leading to low morale and poor performance.[4] They argue that the transformational leadership required in the first revolutionary phase of change must be complemented by a more transitional or transactional leadership style as the organisation moves to institutionalise the changes already begun. In the case of TVNZ, this process was facilitated by the introduction of a more transitional and consultative leadership style in the person of a new chief executive following Mounter's resignation in 1991. By that stage the organisational turnaround had been initiated, a new strategic direction established, new structures put in place and a new entrepreneurial culture introduced.

An important condition of any successful transformation is that people involved with the organisation accept the need for change. These include not only the staff of the organisation but also important external constituencies, which in the case of a television broadcaster include politicians, the press, and the public. To be effective, therefore, leaders of change must have regard for these external forces and pressures. As a highly visible media organisation, external pressures on TVNZ are perhaps greater than they are for most other commercial companies. In this respect at least, TVNZ often encountered difficulties with its external constituencies.

174

Radical Organisational Change

Establishing Clear Strategic Direction

With the deregulation of both broadcasting and the related telecommunications industry, and facing rapid technological change, TVNZ had no choice but to respond to the threats and opportunities of its new environment if it wished to survive. Unlike most organisations, which usually pursue either a defender or prospector strategy,[5] TVNZ developed both simultaneously in response to its new environment. The organisation moved first to protect its existing revenue flows by marshalling its superior resources to defend itself against new entry. It also sought to lower costs and increase productivity in its core businesses of television broadcasting and signal linking and transmission. However, with new competitors hawking new products and forms of delivery, the company recognised that it needed to find new sources of revenue and profits to avoid being outflanked.

But first TVNZ had to ensure that it did not lose too much market share to its new competitor. TVNZ's leadership saw it as critical that it maintain its industry leadership even though it would inevitably lose a percentage of its audience to the opposition. Thus was born the two-channel strategy whereby each of its two channels was branded in such a way as to be complementary to the other and to cover key markets. The intention was to programme each channel so that it attracted different segments of the market, segments that were attractive to advertisers. This strategy was well conceived and implemented, with the consequence that TV3 was unable to break through the ratings into first or second place. As a consequence TVNZ was able to maintain its advertising revenue stream even in the face of competition, thereby providing the company with the strength to pursue other strategic expansion paths.

In order to sustain its defender strategy, a key objective of TVNZ Ltd throughout this period was to retain control of its two channels. In addition to allowing complementary programmes to be offered to New Zealand viewers, TVNZ's managers argued that ownership of two channels provided the synergy and balance sheet strength necessary to finance new technology, to diversify into related activities in electronic communications, and to move into overseas markets.

Not satisfied with its defender two-channel strategy, TVNZ looked for growth opportunities that would complement its existing capabilities and competencies in television broadcasting and electronic signal distribution and transmission. Rather than growing by expanding the boundaries of the firm, TVNZ chose instead to seek economies of scope through diversification alliances.[6] Here TVNZ was acting as a prospector organisation, to use Miles and Snow's terminology again, not content merely to defend territory and position but seeking out strategic expansion paths in related and converging industries.

If it was to replace these lost revenues and grow, managers believed that the company must expand outside its core broadcasting and signal distribution and transmission businesses. Their view was that the company needed to be able to expand into telecommunications and, in the longer run, position the group to market and deliver multi-media services involving television, telephone, and computer technology. To do this, however, TVNZ had to broaden the definition of its core businesses in its Statement of Corporate Intent and gain the agreement of the shareholding ministers to this redefinition. The way was eased by the Rennie report which signalled the need for such a redefinition if TVNZ was to increase its value as a company. This change was accomplished in the negotiation of TVNZ's 1991 Statement of Corporate Intent, where the directors indicated that they aimed 'to operate a commercially successful electronic telecommunications business' rather than 'a commercially successful television business', as had been stated in the 1990 SCI.

TVNZ used this expanded definition of its core businesses to pursue a strategy of related diversification through alliances with major international companies involved in the telecommunications, cable television and entertainment industries. This strategy led to important investments in Sky Network Television and CLEAR Communications. These investments and the strategic alliances which came from them were important to TVNZ for a number of reasons.

First, each alliance provided an opportunity to learn about a related business offering prospects for the transfer of information, skills, and technology.[7] For example, TVNZ's shareholding in Sky provided it with access to knowledge about the operation and management of pay-TV, the fastest growing part of the television broadcasting industry around the world. Similarly, TVNZ's investment in CLEAR opened a window on an array of business opportunities in the burgeoning (and converging) fields of electronic communications of voice, data, and visual images.

TVNZ also investigated the acquisition of a stake in Bond Media Ltd. Although this foray was frustrated by opposition from its shareholders, TVNZ's strategy was once again to form an alliance secured by an equity investment. TVNZ believed that it was strategically important to have investments and relationships throughout the international value chain, i.e., from programme producers through distributors to network operators. In this case TVNZ's strategy also involved a cross-holding by the Australian company in South Pacific Pictures, the objective being to bring them into a supply relationship with TVNZ.

Second, diversification by alliance rather than by changing the boundaries of the firm allowed TVNZ Ltd to expand while economising on limited capital and management capabilities. Wherever possible, TVNZ used its ownership of a complementary asset (its communication and transmission network) both to leverage its investment and to reduce its risk exposure. TVNZ received its stake in CLEAR by allowing CLEAR exclusive access to its network rather than by making a large cash investment. In both CLEAR and Sky, it also reduced its exposure to risk by negotiating contracts for the supply of network or transmission services. In each case TVNZ, with its subsidiary BCL, was both shareholder and supplier.

Third, the alliances themselves were exploited as TVNZ sought to enter new markets. Through its subsidiary BCL the TVNZ group formed a consortium with Bell Canada, another major shareholder in CLEAR, to enter the Australian market to bid on telecommunications contracts. Similarly, the alliance TVNZ joined to operate an Asian financial and business television news service included TeleCommunications Inc. (TCI), one of TVNZ's partners in Sky Television Network Limited.

Alliances with TCI and Time Warner were also important for continued access to major programme sources in the USA. This was particularly critical for Channel 2 which remained heavily dependent on American products for its competitive operation. It was with this long-term view towards protecting its programme supply lines that TVNZ valued its relationships with its alliance partners.

To conclude, TVNZ was not content with defensive strategies but sought out strategic expansion paths by investing in a set of related businesses which complemented and added to its existing capabilities and competencies. Not all of its expansion initiatives were endorsed by its risk-averse owner, but during this period TVNZ became an outward-looking electronic communications company. A major objective of the resulting alliances was not only to increase the utilisation of some core TVNZ activities, but also to provide some protection for programme sources and to open up new growth opportunities. In the future TVNZ was expected to try to capitalise on what it now considered to be its core competencies or skills, which were identified in TVNZ Ltd's *1992 Business Plan* as making, packaging and distributing communications/television products.

Aligning Organisational Structure, Reshaping Employment
Relations, and Changing Organisational Culture
In addition to a clear strategic direction, TVNZ needed mutually supporting organisational structures, processes, and systems to avoid being overwhelmed by new competition and rapid changes in technology. An important challenge for TVNZ was to align its organisational structure and change its culture to provide coherent support for its strategy.

As a first step towards such structural realignment, TVNZ redefined its core businesses and established South Pacific Pictures and Broadcast Communications Ltd as subsidiaries. In so doing, it relocated upstream drama production and downstream transmission facilities outside its core. The Avalon Television Centre and other support activities such as design work and external broadcasting capacity were treated similarly and given devolved responsibility, even though not formally established as subsidiaries. Managers of these units were subject to bottom-line discipline and were accountable to internal governance boards set up on the corporate model. Not all of these units were successful, but the restructuring enabled TVNZ to focus on its core activities and increased the flexibility and competitiveness of its non-core operations.

TVNZ also moved towards a strategic business unit structure within its core operations. A number of SBUs were formed, some as profit centres, others as cost centres. At the same time an internal charging regime, along with the contracting out of some support services, pushed internal costs down and increased TVNZ's competitiveness. Even though requiring modification later, these structural reforms increased internal transparency and accountability, allowed for better performance measurement and helped to change managers' mind-sets and attitudes. Although the SBU structure encountered problems which had to be worked through, it facilitated the new strategic direction of the company and helped institutionalise the changed character of the organisation after corporatisation.

At the same time TVNZ needed to change significantly its human resource policies and industrial relations regime if it was to be competitive and responsive to the market. What was previously a rigid and bureaucratic personnel structure reflecting the old public service model with multiple job designations and classifications, inflexible job demarcations, a multitude of penal rates and special payments and a complex appeal structure was gradually transformed into a more flexible structure. By 1992 most staff were on flat rates and able to work flexible hours without automatically incurring penal rates, managers and supervisory staff were on individual contracts, the work-force was increasingly multi-skilled, there were many fewer job designations, and demarcation disputes were something of the past. In line with the trend in other state-owned enterprises, responsibility for day-to-day human resource management was devolved to line managers thereby increasing management discretion and accountability.[8]

A major objective of any organisational transformation process must be to change the attitudes of managers and staff and to infuse the organisation with a culture which is aligned with strategic direction. Changing the organisational culture is critical to the success of large-scale organisational change. Otherwise, over time the organisation will tend to revert to its previous dominant culture and the change process will be undermined. It is very difficult for organisations with strong histories and cultures, such as a public broadcasting corporation, to change tracks as completely, as was required in this case.[9] Frequently, organisations begin the change process but only get half way or revert to their previous condition. Julian Mounter commented: 'What you see in many other TV companies around the world is a meandering path towards change. Those who meander for too long tend to falter, stumble and occasionally fall.' Encouraging the adoption of a culture that reinforces the new direction of the organisation greatly assists the process of staying on the new track.

Mounter recognised the importance of culture change and invested considerable resources in seeking to instil in the company the new commercial, entrepreneurial spirit that he saw as necessary to cement in the change. Hence the company-wide training programmes which tried to get employees to take an ownership view and to understand the new competitive environment in which TVNZ now operated. Hence Mounter's impassioned speeches littered with metaphors of battle and war. Hence the adoption of a new corporate image and a strong public relations campaign directed not only at external audiences but also at reinforcing staff confidence in the performance of the restructured broadcaster.

However, training programmes, new logos and symbols, and the use of new language and labels are not enough by themselves to make a new culture stick. Appropriate organisation structures, planning and accounting processes, personnel policies and work practices and incentive systems play an important role in reinforcing new ways of thinking and in countering the natural tendency to revert to old practices and attitudes. Culture change is difficult to achieve. It requires the mutual interaction of new symbols and definitions and of changed structures, expectations, and rewards. New attitudes and beliefs need to be demonstrated in new behaviours and expectations.

The extent of cultural change throughout TVNZ four years after its establishment as an SOE undoubtedly varied from one part of the organisation to another. However, TVNZ Ltd in 1992 was a very different organisation from what it had been as a division within the BCNZ in the 1980s. For one thing, it was no longer so production oriented. Indeed, it was clear that an important power shift had occurred within the company from the producers to the programmers. No longer was it enough to make 'good' programmes but it was necessary to convince the programmers that they would bring in the audiences. The image of producers bringing calculators to budget meetings reflects the type of change in attitude that was important to the remaking of TVNZ.

Conclusion

Change has been a feature of public broadcasting systems since the late 1980s. During this time most countries have grappled with issues raised by technological change, increased competition from private broadcasters, and a reluctance of governments to continue to invest in capital developments. Even the august BBC, subject of so much admiration around the world, has embarked on quite radical organisational change under new leadership. Similar concerns drive the BBC's change process as have driven that of TVNZ in New Zealand.[10] It has even been suggested that the BBC could learn something about change from its younger, but more commercially experienced, counterparts such as TVNZ Ltd.[11]

But of all the Anglo-American countries with strong public broadcasting traditions, New Zealand went the furthest in restructuring the entire broadcasting system. Not only were Television New Zealand and Radio New Zealand reformed as state-owned enterprises with clear commercial obligations but a new purchasing authority was established in the form of the Broadcasting Commission. A clear separation was made between commercial and non-commercial objectives, in a manner true to the state-owned enterprise model, with the Broadcasting Commission set up to ensure that the government's social objectives were met. In making this distinction and establishing a new institutional framework to implement it, New Zealand blazed a new trail that has given rise to considerable international interest.[12]

The unique combination of technological, economic, political, social, and cultural factors made the remaking of TVNZ from a division of the BCNZ to TVNZ Ltd both complex and risky. In terms of financial performance and the strategic positioning of TVNZ so as to add value to the government's investment, the transformation has succeeded. TVNZ weathered a period of recession and slow economic growth, and despite suffering the loss of 15–20 percent of its audience and potential advertising revenues, still managed rates of return above those of comparable Australian and US broadcasters.

Five primary drivers of organisational change can be identified. The first was political and social pressure for change. The second was rapid changes in technology and the globalisation of television broadcasting. The third was the deregulation of broadcasting and telecommunication markets and the entry of new competitors. The fourth was the freedom to manage which resulted from TVNZ's establishment as an SOE. The fifth was an experienced and hard-driving chief executive who set and communicated a clear strategic vision and direction for the future of the company. These factors created an internal environment in which managers were able to implement needed changes in a range of areas including the internal structure of the organisation, sales and marketing, human resource management and information, accounting, and financial management systems.

TVNZ survived the introduction of competition when many believed that, as a state-owned broadcaster with an ingrained public service culture, it would surely be overwhelmed. It has since developed and expanded into a substantial, broad-scale company special-ising in electronic and broadcast communications. The combination of TVNZ, Sky, BCL and CLEAR means that TVNZ Ltd is well posi-tioned to capitalise on the current convergence of computers, tele-communications and the visual media.

But at the end of study period, TVNZ remained vulnerable as a broadcasting company. First, its ability to raise capital for investment in new technologies or strategic expansion is severely constrained under Crown ownership. Unable as an SOE to raise equity capital in the share market, TVNZ also operates under strict gearing ratios that limit its debt levels. This places some restrictions on the com-pany's freedom of action.

Second, because of its high profile as New Zealand's major state-owned television broadcaster, TVNZ remains at the centre of politi-cal and social debate about television and its effects. The corporatisation and subsequent commercialisation of TVNZ, with the focus on ratings and audience size that accompanied it, increased the intensity of the debate about the quality and mix of programming being delivered. Cultural critics, in particular, argue that the public should expect 'quality' programmes as one of the dividends from the state-owned broadcaster. As a result the company, although a com-mercial success, remained particularly vulnerable to social and po-litical pressure and to the threat of future regulatory and legislative intervention so long as it remains in public ownership.

To conclude, following corporatisation in late 1988, the remak-ing of TVNZ resulted in an outward-looking, entrepreneurial com-pany which has provided its owner with an improving return on its investment. With strong leadership, improved management processes and systems, an energetic culture and a strategic vision of the future, TVNZ was well positioned at the end of our study in 1992 to take advantage of an economy that was emerging from recession.

Epilogue

The first chief executive of TVNZ Ltd, Julian Mounter, resigned from TVNZ in 1991. After a period as a consultant, he was appointed as chief executive of Star TV in Hong Kong until it was taken over by Rupert Murdoch's News Corporation in 1993. He now works as an independent consultant and director of a number of broadcasting companies in the UK and Europe. Mounter's successor as chief executive was Brent Harman.

Brian Corban's term as the first chairman of TVNZ Ltd ended in late 1992. He is now Chairman of WEL Energy Group Ltd, a member of the Waitangi Tribunal, a company director and a consultant. His successor as chairman of TVNZ was Norman Geary, who had been chief executive of Air New Zealand from 1982 to 1988 and a director of a number of New Zealand companies.

Geary's arrival as chairman and other changes to the board resulted in a shift away from the outward-looking, strategic orientation of TVNZ towards a focus on current profitability. Under Geary's 'hands-on' operational approach, the company reduced its concentration on new technologies and the development of new business areas which the first board and management team saw as so critical to long-term value creation. Instead, increased attention was given to wringing profits from TVNZ's core broadcasting business through further downsizing and cost-cutting. Activities which were not seen to be contributing to current profits were eliminated or curtailed.

One casualty of the refocusing was Asian Business News, a joint venture involving TVNZ, Dow Jones of New York, Tele-Communications Incorporated, and the Singapore Broadcasting Company. Put into place by Brent Harman after almost three years in the planning, this strategic initiative was promoted as an important stepping stone into Asia and the world's fastest growing broadcasting market. However, the project proved to be a drain on company resources and during 1995 the board sold TVNZ's 29.5 percent holding to ABN's other major shareholders. The decision to pull out of the Asian market was reported to have further strained relationships between the chairman, who wished to redirect the company, and the chief executive, who was in favour of a more expansionist, global view of the business. Brent Harman subsequently resigned as chief executive and left TVNZ towards the end of 1995 to take up a position as chief operating officer of a subsidiary of TeleCommunications International in the UK.

After Harman's departure an Australian, Chris Anderson, was appointed as group chief executive. Prior to his appointment he had been managing editor of the Australian Broadcasting Corporation. He had also worked for the Fairfax group, publisher of a number of leading Australian newspapers and a former owner of the Channel 7 network. Darryl Dorrington, TVNZ's deputy chief executive and an unsuccessful applicant for the top job, resigned soon after Anderson's appointment and was followed shortly thereafter by Des Brennan, the director of sales and marketing. The director of programming, John McCready, who is credited with having a major influence on the on-screen visual identity and programming of the two channels after TVNZ was established as an SOE, resigned from the company at the end of 1994. He was replaced by Mike Lattin, another Australian, who came to TVNZ from the Australian Channel Ten Group, where he had been director of network programming.

Ruth Harley, the first executive director of the Broadcasting Commission (New Zealand on Air), resigned from that organisation in 1995. She is now National Media Director for the advertising firm of Saatchi and Saatchi in Wellington.

In 1994 the final transfer of assets to TVNZ Ltd finally took place. Because of an action taken by the Maori Council against the broadcasting assets in 1989, TVNZ had operated its assets under licence from the Crown since its establishment as an SOE. A ruling by the Privy Council in December 1993 in the Crown's favour finally cleared away the last barrier to the television broadcasting assets being vested in the company.

In 1995 TVNZ launched a new network of free-to-air regional stations in Auckland (ATV), Hamilton (Coast to Coast), Wellington (Capital City) and Dunedin (Southern) operating on the UHF frequency. The four stations were to operate independently under the ownership of Horizon Pacific Television, which, in turn, was owned by TVNZ Ltd. Interestingly, this venture was headed by Trevor Egerton, who was TV3's chief executive when it first entered the market in November 1989 to compete against TVNZ. Having been criticised for its decision to exit regional television when it became an SOE, TVNZ was now criticised for returning.[1] By the end of 1995 many households in Auckland had the option of tuning in to up to ten channels including Sky Network Television if they wished. Outside of Auckland almost every New Zealand household had access to three or more channels.

TVNZ's organisation structure has continued to change. In addition to reorganisations that have taken place within major business units such as Avalon and BCL, a major change took place towards the end of 1994 when four units within TVNZ Networks (news and current affairs, sports, operations, and production) were made fully-fledged SBUs with their own boards of management. This move to create an 'internal market' within TVNZ Networks complicated reporting lines, drove up the internal transaction costs of running the company's SBUs, and brought the now enlarged SBU structure under scrutiny. This structure had served the company well by developing a commercial outlook among TVNZ's managers, producers, and staff. However, with the 1994 addition of yet more SBUs, the structure was made more complex and costly to operate. Under considerable pressure from the board to reduce costs and eliminate duplication, the new group chief executive, Chris Anderson, announced a reorganisation of the company soon after his arrival in 1995. The SBU structure was dropped and replaced with a functional structure which included a corporate department and three major divisions: television (operating the two channels, programming, news and current affairs, sport, sales and marketing, enterprises, teletext, eTV, Maori programmes, and Horizon Pacific Television); production (Avalon, South Pacific Pictures, natural history, moving pictures, operations, production, children's programmes, and the New Zealand television archive); and distribution and services (BCL, satellite and Pacific services, and engineering).

Over the 1993–1995 period, the improving trend in profits and dividends discussed in chapter 8 of our study continued. Riding on the crest of a more buoyant economy than had been seen in New Zealand since 1989 and the work done to strengthen the company over the 1989–1992 period, the TVNZ group posted significant improvements in revenues, earnings, dividends, and taxes. By 1995 revenues were up by 25 percent over the 1992 figures of $343.5m to $427.5m. Earnings after tax were up by 84 percent over the 1992 figure of $23.4m to $43m. This suggests that at least part of improvement in current profitability can be attributed to the decisions to concentrate on broadcasting operations, to downsize and to cut costs further within the company. As TVNZ's shares are not traded, it is not easy to determine whether these actions are seen as actually harming or enhancing overall shareholders' value. However, from a purely accounting perspective, the company's return on assets and shareholder funds have increased as have its dividend payments and taxes. Over the 1989–1992 period TVNZ paid $37.7m in dividends to shareholding ministers and taxes of $28.8m. Over the 1993–1995 period TVNZ paid dividends of $85.3m and taxes of $58.7m.

The television market in New Zealand continues to fragment, but TVNZ and TV3 still hold 90–95 percent of the market with Sky Network Television and the regional channels accounting for the remainder. In addition to pay-TV and the new regional channels, Telecom New Zealand has been trialing cable television in parts of Auckland offering 15 channels of viewing. Another cable company, Kiwi Cable Television, is operating in Wellington and on the nearby Kapiti Coast. There has been some recent slippage in TVNZ's audience shares with the greatest dip being in TV2's ratings. Television commentators have speculated that this decline may be due to the reorganisation of the programming unit and the appointment of a new and relatively inexperienced programming team.

Local New Zealand programming during prime time has continued to increase, rising from 1,640 hours in 1992 to 1,821 hours in 1994 for all broadcasters. However, over the same period there has also been some decrease in local content throughout the schedule as a whole. This decrease was spread across all genres with the exception of drama/comedy, news and current affairs, documentaries and information, all of which increased.

The question of whether TVNZ Ltd should remain in Crown ownership is still a topic of current interest.[2] One option that continues to attract attention is to sell Channel 2 to a private operator and to turn Television One into a public broadcaster. Another option, which is favoured by some of the senior executives who have recently departed TVNZ, is to sell TVNZ as a whole.[3] A public-service channel could then be developed using a UHF national network. It is possible that Horizon Pacific Television Limited could fulfil this role. Whether it will be possible for any government to gain sufficient support to sell off all or part of TVNZ with a parliament elected under the new mixed-member proportional voting system now introduced in New Zealand remains an open question.

APPENDIX

Summary of results for the Broadcasting Corporation of New Zealand and Television New Zealand Limited

PANEL A: SUMMARY INCOME STATEMENTS 1979–1992 ($000'S)

TVNZ balance date 31 December as of 1989

	1979	1980	1981	1982	1983	1984
	BCNZ	BCNZ	BCNZ	BCNZ	BCNZ	BCNZ
INCOME	$	$	$	$	$	$
Television advertising	36,504	46,300	55,476	73,944	94,123	98,365
Radio advertising	16,618	17,156	20,930	27,091	32,941	34,598
Listener	5,374	6,890	9,311	11,870	15,236	16,319
Licence fees	29,467	30,694	32,255	33,911	33,863	34,806
Promo's & prog. sales	534	534	772	1,260	2,052	2,754
Commercial productions	644	636	523	630	650	790
Concert proceeds	473	425	738	740	751	1,346
Grants in aid	443	468	180	180		
Investment income	470	562	481	158	945	2,163
Other income	723	784	1,567	1,135	1,399	4,623
TOTAL INCOME	91,250	104,449	122,233	150,919	181,960	195,764
	14.46%	17.03%	23.47%	20.57%	7.59%	15.14%
EXPENDITURE	85,026	95,206	115,523	137,612	156,884	173,003
Growth in expenditure	11.97%	21.34%	19.12%	14.00%	10.27%	9.95%
EBITD	6,224	9,243	6,710	13,307	25,076	22,761
Depreciation	4,849	5,775	6,205	7,044	8,669	9,448
EBIT	1,375	3,468	505	6,263	16,407	13,313
Interest	56	60	58	72	37	47
EBT	1,319	3,408	447	6,191	16,370	13,266
Taxation						
EAT	1,319	3,408	447	6,191	16,370	13,266
Dividend						
Extraordinaries	37	355	233	331	37	399
RETAINED EARNINGS	1,356	3,763	680	6,522	16,407	13,665

KEY: EBITD, Earnings before interest, tax and depreciation; EBIT, Earnings before interest and tax; EBT, Earnings before tax; EAT, Earnings after tax.

1985	1986	1987	1988	1989	1990	1991	1992
BCNZ	BCNZ	BCNZ	BCNZ	TVNZ	TVNZ	TVNZ	TVNZ
$	$	$	$	$	$	$	$
119,909	148,571	168,798	205,988	265,810	226,295	238,259	245,830
37,890	43,275	56,379	61,399				
17,488	18,851	21,037	24,655				
36,103	35,581	41,597	51,222	16,408	17,656	16,439	20,801
4,532	6,276						
830	1,006						
905	1,422	1,873	1,703				
4,525	8,141	3,416	1,293				
3,227	4,069	10,576	13,322	30,443	48,289	59,449	76,832
225,409	267,192	303,676	359,582	312,661	292,240	314,147	343,463
18.54%	13.65%	18.41%	-13.05%	-6.53%	7.50%	9.33%	
190,215	228,989	282,238	310,236	264,209	254,210	256,654	282,702
20.38%	23.25%	9.92%	-14.84%	-3.78%	0.96%	10.15%	
35,194	38,203	21,438	49,346	48,452	38,030	57,493	60,761
11,507	15,142	14,564	17,255	6,511	12,601	14,873	16,951
23,687	23,061	6,874	32,091	41,941	25,429	42,620	43,810
87	89	174	1,139	5,937	13,849	12,952	8,577
23,600	22,972	6,700	30,952	36,004	11,580	29,668	35,233
				9,327	-1,422	9,095	11,837
23,600	22,972	6,700	30,952	26,677	13,002	20,573	23,396
		1,000	6,268	8,000	8,500	11,250	10,000
480	168	675	9,286	-7,045	11,415	7,552	1,396
24,080	23,140	6,375	33,970	11,632	15,917	16,875	14,792

PANEL B: SUMMARY BALANCE SHEETS ($000'S)

TVNZ balance date 31 December as of 1989

	1979	1980	1981	1982	1983	1984
	BCNZ	BCNZ	BCNZ	BCNZ	BCNZ	BCNZ
CURRENT ASSETS	$	$	$	$	$	$
Bank	2,856	6,601	2,539	812	10,561	13,416
Debtors	9,345	11,023	12,079	19,440	24,612	31,270
Rights & inventories	10,753	11,079	15,181	17,654	19,144	22,987
	22,954	28,703	29,799	37,906	54,317	67,673
CURRENT LIABILITIES						
Revenue in advance	145	403	361	652	612	450
Accrued payback	1,328	2,187				
Creditors	7,203	7,289	12,625	13,519	15,377	23,060
Provisions						
Bank overdraft						
	8,676	9,879	12,986	14,171	15,989	23,510
WORKING CAPITAL	14,278	18,824	16,813	23,735	38,328	44,163
INVESTMENTS	5	5				
FIXED ASSETS	53,000	52,217	54,910	54,509	85,552	93,375
FUTURE INC. TAX BENEFIT						
TERM LIABILITIES						
NET ASSETS	67,283	71,046	71,723	78,244	123,880	137,538
SHAREHOLDERS' FUNDS						
Paid-up capital	38,900	38,900	38,900	38,900	38,900	38,900
Capital reserves	25,664	25,664	25,664	25,664	25,664	25,664
Revaluation reserve					29,198	29,191
Retained earnings	2,719	6,481	7,159	13,681	30,087	43,784
FUNDS	67,283	71,045	71,723	78,245	123,849	137,539
TOTAL ASSETS	75,959	80,925	84,709	92,415	139,869	161,048

1985	1986	1987	1988	1989	1990	1991	1992
BCNZ	BCNZ	BCNZ	BCNZ	TVNZ	TVNZ	TVNZ	TVNZ
$	$	$	$	$	$	$	$
28,591	29,105	8,060	0	400	599	1,268	1,692
32,770	42,762	42,916	54,793	45,612	42,611	45,766	46,638
33,634	33,368	34,746	34,491	58,354	73,505	64,571	68,610
94,995	105,235	85,722	89,284	104,366	116,715	111,605	116,940
480	612	949	1,894				
32,149	43,968	47,773	61,939	48,177	45,874	45,872	54,734
			15,500	37,529	16,514	12,469	14,535
			6,220	5,874		690	
32,629	44,580	48,722	85,553	91,580	62,388	59,031	69,269
62,366	60,655	37,000	3,731	12,786	54,327	52,574	47,671
		37	113	11,343	27,111	37,272	45,569
99,253	124,104	154,097	267,171	165,432	173,082	179,241	181,821
					4,879	719	3,836
				37,929	93,960	87,492	81,791
161,619	184,759	191,134	271,015	151,632	165,439	182,314	197,106
38,900	38,900	38,900	38,900	140,000	140,000	140,000	140,000
25,664	25,664	25,664	25,664				
29,191	29,191	29,191	84,334				
67,864	91,004	97,379	122,117	11,632	25,439	42,314	57,106
161,619	184,759	191,134	271,015	151,632	165,439	182,314	197,106
194,248	229,339	239,856	356,568	281,141	321,787	328,837	348,166

PANEL C: RATIO ANALYSIS 1979–1992

TVNZ balance date 31 December as of 1989

	1979	1980	1981	1982	1983	1984
ROE = EAT /SHF	1.96%	4.80%	0.62%	7.91%	13.22%	9.65%
ROA = EBIT/ASSETS	1.81%	4.29%	0.60%	6.78%	11.73%	8.27%
EBIT / EAT	104.25%	101.76%	112.98%	101.16%	100.23%	100.35%
SHF / ASSETS	88.58%	87.79%	84.67%	84.67%	88.55%	85.40%
SALES / ASSETS	120.13%	129.07%	144.30%	163.31%	130.09%	121.56%
EBIDT/SALES	6.82%	8.85%	5.49%	8.82%	13.78%	11.63%
EBIT / SALES	1.51%	3.32%	0.41%	4.15%	9.02%	6.80%
EBITD / ASSETS	8.19%	11.42%	7.92%	14.40%	17.93%	14.13%
EMPLOYEES						3548
TOTAL INC/EMPLOYEES						55.18
TVADVT/.63EMPLOYEES						44.01
EBIT / EMPLOYEES						3.75
EBT / SHF	1.96%	4.80%	0.62%	7.91%	13.22%	9.65%
EBIT / INTEREST	24.55	57.80	8.71	86.99	443.43	283.26
DIV / EAT	0.00%	0.00%	0.00%	0.00%	0.00%	0.00%
GROWTH IN SALES		14.46%	17.03%	23.47%	20.57%	7.59%
GROWTH IN EBIT		152.22%	-85.44%	1140.20%	161.97%	-18.86%
GROWTH IN EXPENSES		11.97%	21.34%	19.12%	14.00%	10.27%

KEY: ROA, Return on assets; ROE, Return on shareholders' funds;
SHF, Shareholders' funds.

APPENDIX

1985	1986	1987	1988	1989	1990	1991	1992
14.60%	12.43%	3.51%	11.42%	17.59%	7.86%	11.28%	11.87%
12.19%	10.06%	2.87%	9.00%	14.92%	7.90%	12.96%	12.58%
100.37%	100.39%	102.60%	103.68%	157.22%	195.58%	207.16%	187.25%
83.20%	80.56%	79.69%	76.01%	53.93%	51.41%	55.44%	56.61%
116.04%	116.51%	126.61%	100.85%	111.21%	90.82%	95.53%	98.65%
15.61%	14.30%	7.06%	13.72%	15.50%	13.01%	18.30%	17.69%
10.51%	8.63%	2.26%	8.92%	13.41%	8.70%	13.57%	12.76%
18.12%	16.66%	8.94%	13.84%	17.23%	11.82%	17.48%	17.45%
3508	3554	3833	3792	2205	2035	1779	1632
64.26	75.18	79.23	94.83	141.80	143.61	176.59	210.46
54.26	66.36	69.90	86.22	120.55	111.20	133.93	150.63
6.75	6.49	1.79	8.46	19.02	12.50	23.96	26.84
14.60%	12.43%	3.51%	11.42%	23.74%	7.00%	16.27%	17.88%
272.26	259.11	39.51	28.17	7.06	1.84	3.29	5.11
0.00%	0.00%	13.56%	15.58%	40.75%	34.81%	40.00%	40.34%
15.14%	18.54%	13.65%	18.41%	-13.05%	-6.53%	7.50%	9.33%
77.92%	-2.64%	-70.19%	366.85%	30.69%	-39.37%	67.60%	2.79%
9.95%	20.38%	23.25%	9.92%	-14.84%	-3.78%	0.96%	10.15%

NOTES

CHAPTER ONE

1 'The Kiwi experiment', *The Economist*, 3 November 1990; 'Return to Rogernomics', *The Economist*, 23 March 1991; 'The mother of all reformers', *The Economist*, 16 October 1991.

2 See Barry Spicer, Robert Bowman, David Emanuel, and Alister Hunt, *The Power to Manage: Restructuring the New Zealand Electricity Department as a State-Owned Enterprise*, Auckland: Oxford University Press, 1991; and Barry Spicer, David Emanuel, and Michael Powell, *Transforming Government Enterprises: Managing Radical Organizational Change in Deregulated Environments*, Sydney: Centre for Independent Studies, 1996.

CHAPTER TWO

1 For the history of television in New Zealand, see R. Boyd-Bell, *New Zealand Television: The First 25 Years*, Auckland: Reed Methuen, 1985.

2 See John Farnsworth, 'Mainstream or Minority: Ambiguities in State or Market Arrangements for New Zealand Television,' in J. Deeks and N. Perry (eds), *Controlling Interests: Business, the State and Society in New Zealand*, Auckland: Auckland University Press, 1992.

3 Some attempts had been made to develop private television broadcasting. Television time had been sold to Northern Television but had proved unsuccessful, and a proposal for breakfast television collapsed when all applicants withdrew.

4 Cameron's minority report was written with the assistance of his son Robert Cameron who was a co-author of R. Cameron and P. Duignan, 'Government Owned Enterprises: Theory, Performance and Efficiency', New Zealand Association of Economists Conference, Wellington, 8 February 1984. This paper had been particularly influential in helping to set the conceptual principles for the reform of government enterprises in New Zealand.

5 Cabinet Minute 88/14/13.

6 It is interesting to note that although Hugh Rennie was chairman of the BCNZ during the period the Rennie Committee prepared its report, it was Brian Corban, the deputy chairman of the BCNZ, who actually chaired the BCNZ board when it formulated the BCNZ's position on restructuring. To avoid a conflict of interest, Rennie took no part in these discussions. The director-general of television at this stage was Julian Mounter. He was later to state that he believed that the separation of television and radio was the best course but because of the internal politics within the BCNZ at the time he did not feel able to lobby for a split away from radio. Instead he argued for a small corporate group of only three or four people and much greater delegation to divisional heads.

CHAPTER THREE

1 Like the chairman, the deputy chairman of the BCNZ was directly appointed by the Minister of Broadcasting under the Broadcasting Act of 1976. The deputy was given certain responsibilities under the Act.

2 Around the time he left the BCNZ in 1988 Nigel Dick sent a paper to the

Rennie Committee in which he advocated (among other things) that Television One be made the 'chartered' broadcaster in television with a lower advertising load, a premium level of local content and a wider range of social service content. In return, Dick advocated that it should receive all the funding going to television, leaving Channel 2 as the unsubsidised commercial competitor to TV3.

3 'Broadcasting Following Deregulation' in Margie Comrie and Judy McGregor (eds), *Whose News?*, Palmerston North: Dunmore Press, 1991.

CHAPTER FOUR

1 The original intention of government was to appoint Hugh Rennie as chairman of the television SOE and Brian Corban as chairman of the radio SOE. However, for professional, family, and philosophical reasons Rennie declined appointment as chairman of TVNZ Ltd.

2 However, it was not long before serious disagreements occurred between TVNZ Ltd and the shareholding ministers and their Treasury advisers over company objectives, dividend levels, profitability targets, and diversification proposals. The development of relationships between TVNZ Ltd and the shareholding ministers and their advisers is discussed in detail in ch. 7.

3 The trade-off was between an extended and conflict-laden valuation and a situation where new board members took longer to get familiar with the business but the relationship between the company and shareholding ministers and their advisers was not damaged in the process. Most importantly the valuation was completed relatively quickly and with a considerable degree of objectivity.

4 Section 9 simply states that nothing in the State-Owned Enterprises Act 1986 'shall permit the Crown to act in a manner that is inconsistent with the Treaty of Waitangi'.

5 Jonathan Hunt has since written that although he was concerned about making state broadcasting services more efficient he also wanted to protect public broadcasting. He wrote 'if there is anything I want to be remembered for in my time as Minister of Broadcasting over nearly 6 years, it is that my aim was always to protect public broadcasting.' See Jonathan Hunt, 'Government and Broadcasting in New Zealand', in G. R. Hawke (ed.), *Access to the Airwaves: Issues in Public Sector Broadcasting*, Wellington, Victoria University Press for the Institute of Policy Studies, 1990.

6 The board members appointed were Brian Corban (chairman) and Janet Clapcott from Auckland, Murray Valentine and Jocelyn Harris from Dunedin, Peter Leeming from Christchurch, Tipene O'Regan from Wellington, and Judy Finn from Upper Moutere. Brian Corban and Tipene O'Regan had been members of the last BCNZ board.

7 These board members were Jocelyn Harris, Peter Leeming, and Judy Finn.

8 For a more detailed discussion of changes in employment practices subsequent to corporatisation see Michael Powell and Barry Spicer, 'The Transformation of Employment Relations in New Zealand State-Owned Enterprises: The Assertion of Management Control', *Asia Pacific Journal of Human Resources*, 32 (2) 1994.

CHAPTER FIVE

1 TV3 original schedules prepared for the Broadcasting Tribunal hearings

placed their news at 6.00 p.m. to be followed by a half-hour talk show. The
view of one senior TV3 manager at that time was that this was done by TVNZ
as a direct response to the TV3 schedule put before the Broadcasting Tribunal during the hearings for a third channel warrant.

2 For benchmarking purposes TVNZ's management viewed Television One as
having some similarities to BBC1 and ABC in Australia with Channel 2 having similarities with Channel Nine in Australia or with wholly commercial
channels.

3 As reported in Television Planning Department, *New Zealand and the International Television Industry*, Television New Zealand Limited, January 1991,
p. 37.

4 In respect of most of the major international organisations like the BBC and
CBS, long-term output contracts were obtained prior to the establishment
of TVNZ Ltd as an SOE.

5 Under an output deal, a broadcaster has first option on a distributor's annual supply of programmes. The broadcaster agrees to take a minimum
number of hours, or in some cases to take programmes worth a minimum
amount. The competitive advantage of output deals with major distributors
such as the BBC and Columbia is access to a continuing stream of popular
programming and the ability to deny these programmes to competitors. The
risk associated with too many output deals is that a broadcaster becomes
swamped with more programmes than needed, yet for which the broadcaster
must pay.

6 With large annual commitments for the purchase of overseas programmes,
TVNZ was subject to risk of movements in the exchange rate. Since establishment as an SOE, the company has operated treasury procedures using
forward cover to protect itself against major movements in exchange rates.

7 Interestingly, on first sighting TV3's schedule when it was released to the
press and advertising agencies, TVNZ Ltd's programmers considered it so bi-
zarre they initially treated it as a blind designed as a tactic to throw them off.

8 A review of TV3's prospectus shows that on the day on which it planned to
commence broadcasting, its cash resources were $3,005,940. However, we
understand that $3m of this amount was in an escrow account to protect
the interest payments to TV3's principal financier, Westpac Bank.

9 TV3 sales and marketing entered the market using the system of graduated
rates and pre-emption that TVNZ had abandoned. Cindy Mitchener, who
was head-hunted from TVNZ Ltd and joined TV3 as sales and marketing
director during their first year on air, quickly dropped the pre-empt system
which had been put into place by people who had left TVNZ and gone over
to TV3 during the start-up period.

10 TV3 alleged that TVNZ Ltd was, in fact, offering 'exclusive' or 'nearly exclusive' deals to advertisers that breached sections 27 and 36 of the Commerce Act 1986. These sections prohibit arrangements that are likely to
substantially lessen competition in a market and forbid a person with a
dominant position in a market from using that position to restrict others
from entering the market or from engaging in competition. This is one of
the grounds cited in an action brought by TV3 Network Ltd against Television New Zealand Limited. This (amended) action was filed in the High
Court in Auckland on 5 December 1991.

11 For the classic development of the value chain concept see Michael Porter,

The Competitive Advantage of Nations, New York: Free Press, 1990.

12 Until 1991 a diary system was used to measure ratings. This was replaced with a computerised system of 'people meters' which are attached to television sets and video recorders in a statistically based sample of 440 homes throughout New Zealand. Minute-by-minute viewing by each member of the household is recorded and the data retrieved nightly by a central computer at AGB McNair. The data is collated into 15-minute audiences broken down by key demographics such as sex and age grouping. The feedback is fast and accurate.

13 Attitudinal data was generated by 'System 7' research. This system was first introduced in the New Zealand radio market to measure the popularity of songs for programming purposes, and later adapted for use in television at Channel Seven in Australia by its New Zealand developer.

14 The positioning strategy and promotion of Television One and Channel 2 was highly successful but it was also very expensive if, in addition to the direct cost, the value of the time-spots utilised for promotion purposes, which might otherwise have been used for advertising, is counted.

15 According to TVNZ's 1989 Business Plan, projected operating costs (excluding depreciation, interest and taxes) were about 85 percent of revenues.

16 TV3's news operation was much smaller that TVNZ's which also produced *Te Karere* (the Maori news programme). However, TV3's news did not rate as highly.

17 Decisions to downsize at Broadcasting Communications Limited (BCL) and Avalon were made by the managers of these units.

18 The valuation and establishment of BCL was not completed by 1 December 1988 because of uncertainty about the pricing of its services to internal and external clients. The Ministerial Advisory Committee was reconvened in 1989 and the same procedure used to set a valuation for TVNZ Ltd as a whole was used to establish a valuation and a financial structure for BCL.

19 On 15 February 1989 TVNZ's board approved an investment of $2.5 million in Sky Network Television Limited with an option to invest another $4.9 million to acquire a 35 percent holding.

20 Currently available digital compression technology offers the potential to split each of Sky's current channels into 3 or 4 channels for a possible 12 to 16 channels in the future. With Telecom studying the feasibility of cable television with a trial in the Auckland area, further fragmentation of the viewing audience is inevitable.

21 The economics of a pay-TV operation are unusual insofar as each new subscriber requires installation of a costly decoder. This results in heavy and risky investment by pay-TV companies in those years in which the subscriber base is being built. It is only later when a substantial subscriber base is established and cash inflows from subscriber income exceeds cash outflows for decoders and installation expenses that a pay-TV operation will achieve positive cash flows.

22 Bell Atlantic and Ameritech were also shareholders of Telecom Corporation of New Zealand.

23 TVNZ Ltd reported a extraordinary gain of $1.55 million on the sale of shares in Sky in 1990 and a further gain of $4 million from a further sell down in 1991.

24 It was originally intended that New Zealand Railways Corporation would

take a minority shareholding. Instead Railways decided to sell a portion of its fibre optic cable to the new company, retaining an option to take a shareholding in partial payment for the cable at a later date.

25 The primary impediment to CLEAR's development was the lengthy negotiation of the terms and conditions under which CLEAR would have access to Telecom's local telephone network. While an interconnection agreement was put into place to allow CLEAR to get into business, CLEAR regarded Telecom's pricing and terms of access as onerous. CLEAR finally filed a High Court action to gain a resolution of this and other competition issues.

26 R. E. Miles and C. C. Snow, *Organizational Strategy, Structure, and Process*, New York: McGraw-Hill, 1978.

CHAPTER SIX

1 Internal contracts and charging also exist in other areas. For example, TVNZ Networks contracts with BCL to provide linking and transmission services. These contracts provide for charges for services, with penalties to be paid by BCL for faults resulting in lost revenues. These penalties provide an incentive to maintain the quality and reliability of its service.

2 See 'Tuning to the Future', *The Economist*, 5 September 1992; and Michael Starks, 'Producer Choice in the BBC', in A. Harrison (ed.), *From Hierarchy to Contract*, London: Policy Journals, 1993.

3 What industry-specific information systems TVNZ, as a division of the BCNZ, had been able to install to support unique activities in the television broadcasting business had been developed despite opposition from BCNZ's central computing operation. Good computing systems, particularly in the specialised TV fields of sales and marketing, TV programming, newsroom automation and resources management were now viewed as a critical success factor in a competitive environment.

4 Some movie rights are bought in bundles. Distributors will often bundle movies and in order to acquire a particular movie it is necessary to buy the bundle in which it is packaged. Most of the cost is loaded onto the movies judged to have the greatest earning potential with a nominal value being placed on the others for accounting purposes.

CHAPTER SEVEN

1 This was a continuation of the tension that had existed throughout 1988 between those (including the Rennie Committee) who wished to see a commitment to the principles of public broadcasting maintained by the new broadcasting SOEs and Treasury officials who felt that social objectives should be excluded and handled in a different way.

2 On 27 February 1990 Treasury officials advised shareholding ministers that TVNZ's proposed target was still below the rate of return that officials believed private investors would require before investing in a television company.

3 TVNZ's financial and accounting systems went through a period of enormous change during the 1990 financial year. This involved internal accounting and budgeting changes necessitated by the restructuring of the company into strategic business units, changes in the methods of costing and writing off programmes and the revision of accounting policies adopted for the New Zealand Listener and Sky Network Television.

4 Section 18 places an obligation on an SOE to supply information on its affairs (subject to certain exclusions) as requested by shareholding Ministers.
5 The change in accounting policy involved a change in accounting for programme stocks. Production overheads, which for local productions had been written off as period expenses, would now be treated as part of the inventoriable cost of local programme production. Unfortunately, although TVNZ had signalled in its 1990 SCI that it intended to change its accounting policy, the impact of the change on targets was not estimated or communicated to Treasury monitors who had difficulty meaningfully comparing the quarterly financial reports of the company with the SCI targets.
6 The complete statement of the policy was that:
a) SOEs are not a vehicle whereby the Government invests to build up a diversified business portfolio;
b) SOEs have a responsibility to maximise shareholders' wealth by efficiently managing the core business and any non-core assets. The government wishes SOEs to focus their investment activities within the core business as defined by the SCI;
c) where an SOE wishes to invest outside the core business as defined in the SCI to add value to the core business, this should not be prohibited; and
d) where there are existing non-core assets which can realise value for the shareholder, the Government's preference is that this be done by means other than direct investment, for example through leasing or licensing agreements. However, it remains the responsibility of the Board to determine the most appropriate way to maximise value for the Crown.
7 Avalon was not legally a subsidiary, although there was an intention at this time to establish it as one.
8 The objective was of course to gain a co-production partner and to gain access to the Australian market for SPP's productions.
9 TVNZ had asked the Crown to provide a repurchase guarantee to purchasers of subsidiary company shares in the event that TVNZ and its subsidiary companies never received their assets from the Crown and were forced to continue to operate them under licence.
10 Australia finally closed the door by lowering the level of foreign ownership allowed in Australian broadcasters.
11 As discussed in chapter 2, the Rennie Committee report recognised that the structure and changes it proposed for TVNZ implied a high-risk strategy for the government to manage and that the shareholders' natural tendency was to take a conservative approach. Indeed, the report noted that the normal preference of the shareholder for a low-risk strategy of retrenchment could lead to a potential tension between the shareholder and TVNZ management with the shareholder trying to restrain TVNZ's high-risk and overseas investments. This is exactly what happened.
12 According to the Australian Broadcasting Tribunal's Broadcasting Financial Yearbook the commercial television industry in Australia experienced overall losses of $81m in 1989/90 and $225m in 1990/91. The Channel Nine network was the only television network to improve its profit before interest and tax in the major markets of Sydney, Melbourne, Brisbane, and Adelaide. Its profits were up from $15.9m in 1989/90 to $49.8m in 1990/91. It is relevant to note that in November 1989 Quintex Australia, the parent company of Channel 7, went into liquidation and that in September 1990,

Northern Star Holding, the parent company of Channel 10, went into receivership.

13 Stan Rodger, Clive Matthewson, Roger Douglas, and David Caygill in the 1987–90 Labour government and Warren Cooper and Doug Kidd in the 1990 National government acted as shareholding ministers. During its term the Labour government specifically removed television and radio from the SOE portfolio and placed them under a separate minister. After Richard Prebble was removed from Cabinet in November 1988, Stan Rodger became Minister for State-Owned Enterprises and Minister Responsible for Television New Zealand Ltd and Radio New Zealand Ltd. To make things even more confused, Jonathan Hunt, as Minister of Broadcasting, was delegated powers by Stan Rodger to deal with some matters that normally would have been dealt with by the shareholding minister. Towards the end of the Labour government's term of office Clive Matthewson was appointed as the Minister Responsible. David Caygill was Minister of Finance and the other shareholding minister throughout this period. When the National government came to power in 1990, Warren Cooper was appointed as the Minister Responsible and Doug Kidd, as Associate Minister of Finance, had delegated authority from the Minister of Finance to act as the other shareholding minister. Maurice Williamson was appointed as Minister of Broadcasting. During the 1988–1992 period there had also been a number of changes in the Treasury officials charged with advising the shareholding ministers on TVNZ Ltd.

14 Perhaps the reason that these terms do not appear in the Act is that they are elusive concepts on which there are a number of different opinions. For example, some people equate 'public service' and 'quality' in television. Some see these things as involving no advertisements, while others stress the importance of particular types of programmes such as thoughtful news and current affairs, serious drama and documentaries, televised ballet and orchestral performances, and programmes for minorities, etc. Some people stress the need for an international perspective and wish to bring more of the rest of the world to New Zealand, while others believe that New Zealanders need to see more of their own world on television. 'Public service' or 'quality' television may also mean different combinations of these things to different people.

15 The Minister believed that ordinary programmes such as *The Billy T. James Show* should receive some sort of public funding because every New Zealander with a television must pay the public broadcasting fee. In his opinion it was wrong to direct the fee only to the '1% or 2% who only watch esoteric programming on television'. See Jonathan Hunt, 'Government and Broadcasting in New Zealand', in G. R. Hawke (ed.), *Access to the Airwaves: Issues in Public Sector Broadcasting*, Wellington: Victoria University Press for the Institute of Policy Studies, 1990. Interestingly, Hugh Rennie recalls that in negotiations for a Maori television project (Aotearoa Broadcasting System) *The Billy T. James Show* was represented as a Maori show and one which would be expected to screen on a Maori television channel.

16 This does not mean that TVNZ Ltd is forbidden to make or screen local or minority-interest programmes and/or to undertake other activities which may not be profit-making. What it does mean is that the decision to undertake such activities must be made in a commercial way and must be com-

mercially sensible within the context of their longer-term strategy and overall profit position.

17 It was not until 1987 that annual reports of the BCNZ started to provide a brief summary of how the public broadcasting fee was spent.

18 The annual public broadcasting fee has failed to keep pace with the movement in the Consumer Price Index. Had the fee kept pace with movements in the CPI, starting from a base of $45 in 1975, in 1990 the fee would have been $269 (exclusive of GST), whereas in fact it was still set at $97.78 (exclusive of GST).

19 The Broadcasting Commission reports for a financial year running from 1 July to 30 June of the following year.

20 Commentators such as Alan Cocker in his occasional series of commentaries on television broadcasting in the *New Zealand Herald* and Joe Atkinson of the Political Studies Department of the University of Auckland have been prominent critics of the current structure of television broadcasting and TVNZ programming. They have been particularly vocal about the quality of news and current affairs programmes.

21 The presentation of a range of opinions on this and related issues covered in this section is given in more detail in G. R. Hawke (ed.), *Access to the Airwaves: Issues in Public Sector Broadcasting*, Wellington: Victoria University Press for the Institute of Policy Studies, 1990.

22 Soon after the 1990 election the Minister of Broadcasting had a paper prepared for the National government's caucus. This paper advanced the argument that Television One with its mixed funding (advertising, sponsorship, and licence fee) and its programming emphasis on educational and cultural information, already had many of the so called 'non-commercial' characteristics sought by National's election policy. The paper pointed out that many of the options for increasing the non-commercial character of Television One involved high costs, or risked undermining the efficiency of broadcasting institutions and limiting the audience appeal of Television One.

23 The relative costs of local production versus acquisition of overseas programme rights were discussed in chapter 5.

24 'Local content grows 42 per cent', *New Zealand Herald*, 16 June 1993.

25 Board minutes show that board members often expressed their views particularly on news and current affairs programmes but did not direct the chief executive on these matters as editor-in-chief.

26 The concern of the MAC echoed those that a range of parties, including ministers and other members of parliament on the government and opposition benches, had raised with officials.

27 TVNZ Ltd's SCIs included the following statement:
'By virtue of its role as a broadcaster Television New Zealand Limited recognises the special requirements for complete and visible independence from Government on all matters relating to programming, editorial standards and personnel matters and will strenuously protect its reputation for independence and impartiality.'
The letters of appointment of the initial board members contained the following statement:
'An important question raised with us by the Ministerial Advisory Committee has been the potential for future governments to dismiss or seek the resignation of Directors for editorial reasons. We would like to assure

you that our intention will be to hold Directors accountable only in terms of the matters laid down in the State-Owned Enterprises Act.'
TVNZ Ltd's licence agreement contained the following statement:
'The parties agree and declare that nothing in this Agreement shall be deemed or construed in such manner as to permit any compromise of the programming independence and editorial integrity of the company.'

28 TVNZ Planning Department, *New Zealand and the International Television Industry*, Television New Zealand Limited, January 1991.

29 Nick Bell, 'The Top 100 Television Companies', *Television Business International*, March 1992, 31 and 34.

30 The analysis reviews BBC2 and Channel 4 (UK) programmes and audience sizes and makes some comparisons with New Zealand. First, most of the top programmes on these two channels were being shown on either TVNZ or TV3. Second, outside the top 20 programmes, the average audience on the two UK channels was less than 5 percent of total homes. By extension, a Channel 4 or BBC2 type channel in New Zealand could expect to reach an audience of around 4 percent of total homes, with average viewing of 39,200 people. It would be difficult to justify television either by licence fee or advertising if it produces average audiences of only 39,200 people.

31 In fact Packer's organisation undertook a due diligence investigation of TV3 as a possible purchase. At this time Julian Mounter requested a $15m fighting fund from TVNZ Ltd's board to be used to defend the company against what would have been a formidable competitor.

CHAPTER EIGHT

1 Those data suggest that television activities comprise about 75 percent of the asset base of the BCNZ, about 77 percent of the combined radio and television revenues (excluding the PBF) and about 63 percent of all staff (including all Broadcasting Services staff in the television numbers).

2 The decision to cease equity accounting for the *New Zealand Listener* resulted from the sale of TVNZ's 50 percent interest in the New Zealand Listener Limited and the acquisition of a 15 percent interest in the New Zealand Listener (1990) Limited which purchased all the New Zealand Listener assets. The decision to no longer equity account for Sky Network Television Limited resulted from the substantial reduction of TVNZ's shareholding. TVNZ's holding in CLEAR was reclassified to an investment because the company claims it does not have significant influence over its operating, financial, and dividend policies.

3 These profits have arisen primarily through the sale of investments in associate companies offset by a provision for restructuring in 1989. One sale contract requires BCL to provide network capacity. If this is not provided part of the profit proceeds from the sale were to be refunded.

4 TVNZ Ltd commenced operations on 1 December 1988. The financial year was shifted to the calendar year (1 January – 31 December) thereafter creating a 13-month reporting period in the first year.

5 The ratios for the BCNZ era are only very rough estimates. They are based on the assumption that 63 percent of the BCNZ's employees were involved in television-related activities.

6 TVNZ Ltd is not the only SOE to exhibit this 'hockey-stick' pattern in its profitability projections. It seems to be a relatively common phenomenon.

7 We eliminated some extreme cases (the highest two RoEs and the six lowest RoEs), as average values can be sensitive to outliers.

CHAPTER NINE

1 Noel M. Tichy and David O. Ulrich, 'The Leadership Challenge—A Call for the Transformational Leader', *Sloan Management Review*, Fall 1984.

2 For a discussion of the role of change agents, see M. Patrickson and G. Bamber, 'Introduction', in M. Patrickson, V. Bamber and G. Bamber (eds), *Organizational Change Strategies: Case Studies of Human Resource and Industrial Relations Issues*, Melbourne: Longman, 1995.

3 D. Stace and D. Dunphy, *Beyond the Boundaries: Leading and Recreating the Successful Enterprise*, Sydney: McGraw-Hill, 1994, 103.

4 Stace and Dunphy, 104.

5 R. E. Miles and C. C. Snow, *Organizational Strategy, Structure, and Process*, New York: McGraw-Hill, 1978.

6 For an interesting discussion of the use of diversification alliances rather than expanding the boundaries of the firm to seek economies of integration, scope and scale, see Torger Reve, 'The Firm as a Nexus of Internal and External Contracts', in M. Aoki, G. Gustafsson and O. E. Williamson (eds), *The Firm as a Nexus of Treaties*, London: Sage Publications, 1990. TVNZ provides an interesting case study of a number of the types of alliances discussed by Reve.

7 For a discussion of the organisation learning that can occur as a consequence of strategic alliances see Gary Hamel, Yves Doz, and C. K. Prahalad, 'Colloborate with Your Competitors and Win', *Harvard Business Review*, January–February 1989.

8 Michael Powell and Barry Spicer, 'The Transformation of Employment Relations in New Zealand State-Owned Enterprises: The Assertion of Management Control', *Asia-Pacific Journal of Human Resources*, 32 (2), 1994.

9 For a discussion of the difficulty in achieving large-scale organisational change, or of jumping from one set of organisational tracks to another, as was required here, see R. Hinings and R. Greenwood, *The Dynamics of Strategic Change*, Oxford: B. Blackwell, 1988.

10 'Reforming the BBC', *The Economist*, 16 March 1991; 'Tuning in to the future', *The Economist*, 5 September 1992; 'The future of the BBC', *The Economist*, 9 July 1994.

11 'Educating Auntie', *Broadcast*, 8 Feb 1991. The establishment of TVNZ as an SOE has also been used as a case study of a public broadcaster in transition by the European Broadcasters' Executive Development Programme.

12 A similar proposal for a purchaser/provider split was floated in a British government green paper on reforming the BBC. See 'Changing Channels', *The Economist*, 28 November 1992.

EPILOGUE

1 'Behemoth of State TV', *New Zealand Herald*, 26 May 1995.

2 'Two Channels to Sell', *New Zealand Herald*, 21 January 1994; 'Channel 2 sale defies new environment', *National Business Review*, 28 February 1994.

3 'Sell both channels, says departing TVNZ boss', *National Business Review*, 28 July 1995; 'TVNZ deputy heads for Japan', *New Zealand Herald*, 2 December 1995.

INDEX

INDEX

205

INDEX